Paul R. Halmos

Finite-Dimensional Vector Spaces

T0351083

Springer

Paul R. Halmos
Department of Mathematics
Santa Clara University
Santa Clara, CA 95053-0290
phalmos@scuacc.scu.edu

Mathematics Subject Classifications (2000): 15-01, 15A03

Library of Congress Cataloging in Publication Data

Halmos, Paul Richard, 1916–
   Finite-dimensional vector spaces.

(Undergraduate texts in mathematics)
Reprint of the 2d ed. published by Van Nostrand,
Princeton, N.J., in series: The University series
in undergraduate mathematics.
   Bibliography: p.
   1. Vector spaces.   2. Transformations (Mathematics).
I. Title.
[QA186.H34 1974]   512´.523   74-10688

Printed and bound by Edwards Brothers, Ann Arbor, Michigan
Printed in the United States of America.

9 8

ISBN 0-387-90093-4
ISBN 3-540-90093-4   SPIN 10875928

Springer-Verlag   New York Berlin Heidelberg
A member of BertelsmannSpringer Science+Business Media GmbH

PREFACE

My purpose in this book is to treat linear transformations on finite-dimensional vector spaces by the methods of more general theories. The idea is to emphasize the simple geometric notions common to many parts of mathematics and its applications, and to do so in a language that gives away the trade secrets and tells the student what is in the back of the minds of people proving theorems about integral equations and Hilbert spaces. The reader does not, however, have to share my prejudiced motivation. Except for an occasional reference to undergraduate mathematics the book is self-contained and may be read by anyone who is trying to get a feeling for the linear problems usually discussed in courses on matrix theory or "higher" algebra. The algebraic, coordinate-free methods do not lose power and elegance by specialization to a finite number of dimensions, and they are, in my belief, as elementary as the classical coordinatized treatment.

I originally intended this book to contain a theorem if and only if an infinite-dimensional generalization of it already exists. The tempting easiness of some essentially finite-dimensional notions and results was, however, irresistible, and in the final result my initial intentions are just barely visible. They are most clearly seen in the emphasis, throughout, on generalizable methods instead of sharpest possible results. The reader may sometimes see some obvious way of shortening the proofs I give. In such cases the chances are that the infinite-dimensional analogue of the shorter proof is either much longer or else non-existent.

A preliminary edition of the book (Annals of Mathematics Studies, Number 7, first published by the Princeton University Press in 1942) has been circulating for several years. In addition to some minor changes in style and in order, the difference between the preceding version and this one is that the latter contains the following new material: (1) A brief discussion of fields, and, in the treatment of vector spaces with inner products, special attention to the real case. (2) A definition of determinants in invariant terms, via the theory of multilinear forms. (3) Exercises.

The exercises (well over three hundred of them) constitute the most significant addition; I hope that they will be found useful by both student

v

and teacher. There are two things about them the reader should know. First, if an exercise is neither imperative ("prove that . . .") nor interrogative ("is it true that . . . ?") but merely declarative, then it is intended as a challenge. For such exercises the reader is asked to discover if the assertion is true or false, prove it if true and construct a counterexample if false, and, most important of all, discuss such alterations of hypothesis and conclusion as will make the true ones false and the false ones true. Second, the exercises, whatever their grammatical form, are not always placed so as to make their very position a hint to their solution. Frequently exercises are stated as soon as the statement makes sense, quite a bit before machinery for a quick solution has been developed. A reader who tries (even unsuccessfully) to solve such a "misplaced" exercise is likely to appreciate and to understand the subsequent developments much better for his attempt. Having in mind possible future editions of the book, I ask the reader to let me know about errors in the exercises, and to suggest improvements and additions. (Needless to say, the same goes for the text.)

None of the theorems and only very few of the exercises are my discovery; most of them are known to most working mathematicians, and have been known for a long time. Although I do not give a detailed list of my sources, I am nevertheless deeply aware of my indebtedness to the books and papers from which I learned and to the friends and strangers who, before and after the publication of the first version, gave me much valuable encouragement and criticism. I am particularly grateful to three men: J. L. Doob and Arlen Brown, who read the entire manuscript of the first and the second version, respectively, and made many useful suggestions, and John von Neumann, who was one of the originators of the modern spirit and methods that I have tried to present and whose teaching was the inspiration for this book.

P. R. H.

CONTENTS

CHAPTER I

SPACES

═══════════════════════════════════════

§ 1. Fields

In what follows we shall have occasion to use various classes of numbers (such as the class of all real numbers or the class of all complex numbers). Because we should not, at this early stage, commit ourselves to any specific class, we shall adopt the dodge of referring to numbers as *scalars*. The reader will not lose anything essential if he consistently interprets scalars as real numbers or as complex numbers; in the examples that we shall study both classes will occur. To be specific (and also in order to operate at the proper level of generality) we proceed to list all the general facts about scalars that we shall need to assume.

(A) To every pair, α and β, of scalars there corresponds a scalar $\alpha + \beta$, called the *sum* of α and β, in such a way that
    (1) addition is commutative, $\alpha + \beta = \beta + \alpha$,
    (2) addition is associative, $\alpha + (\beta + \gamma) = (\alpha + \beta) + \gamma$,
    (3) there exists a unique scalar 0 (called *zero*) such that $\alpha + 0 = \alpha$ for every scalar α, and
    (4) to every scalar α there corresponds a unique scalar $-\alpha$ such that $\alpha + (-\alpha) = 0$.

(B) To every pair, α and β, of scalars there corresponds a scalar $\alpha\beta$, called the *product* of α and β, in such a way that
    (1) multiplication is commutative, $\alpha\beta = \beta\alpha$,
    (2) multiplication is associative, $\alpha(\beta\gamma) = (\alpha\beta)\gamma$,
    (3) there exists a unique non-zero scalar 1 (called *one*) such that $\alpha 1 = \alpha$ for every scalar α, and
    (4) to every non-zero scalar α there corresponds a unique scalar α^{-1} $\left(\text{or } \dfrac{1}{\alpha}\right)$ such that $\alpha\alpha^{-1} = 1$.

(C) Multiplication is distributive with respect to addition, $\alpha(\beta + \gamma)$ $= \alpha\beta + \alpha\gamma$.

If addition and multiplication are defined within some set of objects (scalars) so that the conditions (A), (B), and (C) are satisfied, then that set (together with the given operations) is called a *field*. Thus, for example, the set Q of all rational numbers (with the ordinary definitions of sum and product) is a field, and the same is true of the set $\mathcal{R}$ of all real numbers and the set C of all complex numbers.

EXERCISES

1. Almost all the laws of elementary arithmetic are consequences of the axioms defining a field. Prove, in particular, that if $\mathfrak{F}$ is a field, and if α, β, and γ belong to $\mathfrak{F}$, then the following relations hold.
  (a) $0 + \alpha = \alpha$.
  (b) If $\alpha + \beta = \alpha + \gamma$, then $\beta = \gamma$.
  (c) $\alpha + (\beta - \alpha) = \beta$. (Here $\beta - \alpha = \beta + (-\alpha)$.)
  (d) $\alpha \cdot 0 = 0 \cdot \alpha = 0$. (For clarity or emphasis we sometimes use the dot to indicate multiplication.)
  (e) $(-1)\alpha = -\alpha$.
  (f) $(-\alpha)(-\beta) = \alpha\beta$.
  (g) If $\alpha\beta = 0$, then either $\alpha = 0$ or $\beta = 0$ (or both).

2. (a) Is the set of all positive integers a field? (In familiar systems, such as the integers, we shall almost always use the ordinary operations of addition and multiplication. On the rare occasions when we depart from this convention, we shall give ample warning. As for "positive," by that word we mean, here and elsewhere in this book, "greater than or equal to zero." If 0 is to be excluded, we shall say "strictly positive.")
  (b) What about the set of all integers?
  (c) Can the answers to these questions be changed by re-defining addition or multiplication (or both)?

3. Let m be an integer, $m \geq 2$, and let Z_m be the set of all positive integers less than m, $Z_m = \{0, 1, \cdots, m - 1\}$. If α and β are in Z_m, let $\alpha + \beta$ be the least positive remainder obtained by dividing the (ordinary) sum of α and β by m, and, similarly, let $\alpha\beta$ be the least positive remainder obtained by dividing the (ordinary) product of α and β by m. (Example: if $m = 12$, then $3 + 11 = 2$ and $3 \cdot 11 = 9$.)
  (a) Prove that Z_m is a field if and only if m is a prime.
  (b) What is -1 in Z_5?
  (c) What is $\frac{1}{3}$ in Z_7?

4. The example of Z_p (where p is a prime) shows that not quite all the laws of elementary arithmetic hold in fields; in Z_2, for instance, $1 + 1 = 0$. Prove that if $\mathfrak{F}$ is a field, then either the result of repeatedly adding 1 to itself is always different from 0, or else the first time that it is equal to 0 occurs when the number of summands is a prime. (The *characteristic* of the field $\mathfrak{F}$ is defined to be 0 in the first case and the crucial prime in the second.)

5. Let $Q(\sqrt{2})$ be the set of all real numbers of the form $\alpha + \beta \sqrt{2}$, where α and β are rational.
(a) Is $Q(\sqrt{2})$ a field?
(b) What if α and β are required to be integers?

6. (a) Does the set of all polynomials with integer coefficients form a field?
(b) What if the coefficients are allowed to be real numbers?

7. Let $\mathfrak{F}$ be the set of all (ordered) pairs (α, β) of real numbers.
(a) If addition and multiplication are defined by

$$(\alpha, \beta) + (\gamma, \delta) = (\alpha + \gamma, \beta + \delta)$$

and

$$(\alpha, \beta)(\gamma, \delta) = (\alpha\gamma, \beta\delta),$$

does $\mathfrak{F}$ become a field?
(b) If addition and multiplication are defined by

$$(\alpha, \beta) + (\gamma, \delta) = (\alpha + \gamma, \beta + \delta)$$

and

$$(\alpha, \beta)(\gamma, \delta) = (\alpha\gamma - \beta\delta, \alpha\delta + \beta\gamma),$$

is $\mathfrak{F}$ a field then?
(c) What happens (in both the preceding cases) if we consider ordered pairs of complex numbers instead?

§ 2. Vector spaces

We come now to the basic concept of this book. For the definition that follows we assume that we are given a particular field $\mathfrak{F}$; the scalars to be used are to be elements of $\mathfrak{F}$.

DEFINITION. A *vector space* is a set $\mathcal{V}$ of elements called *vectors* satisfying the following axioms.

(A) To every pair, x and y, of vectors in $\mathcal{V}$ there corresponds a vector $x + y$, called the *sum* of x and y, in such a way that
(1) addition is commutative, $x + y = y + x$,
(2) addition is associative, $x + (y + z) = (x + y) + z$,
(3) there exists in $\mathcal{V}$ a unique vector 0 (called the *origin*) such that $x + 0 = x$ for every vector x, and
(4) to every vector x in $\mathcal{V}$ there corresponds a unique vector $-x$ such that $x + (-x) = 0$.

(B) To every pair, α and x, where α is a scalar and x is a vector in $\mathcal{V}$, there corresponds a vector αx in $\mathcal{V}$, called the *product* of α and x, in such a way that
(1) multiplication by scalars is associative, $\alpha(\beta x) = (\alpha\beta)x$, and
(2) $1x = x$ for every vector x.

(C) (1) Multiplication by scalars is distributive with respect to vector addition, $\alpha(x + y) = \alpha x + \alpha y$, and

(2) multiplication by vectors is distributive with respect to scalar addition, $(\alpha + \beta)x = \alpha x + \beta x$.

These axioms are not claimed to be logically independent; they are merely a convenient characterization of the objects we wish to study. The relation between a vector space $\mathcal{V}$ and the underlying field $\mathcal{F}$ is usually described by saying that $\mathcal{V}$ is a vector space *over* $\mathcal{F}$. If $\mathcal{F}$ is the field $\mathcal{R}$ of real numbers, $\mathcal{V}$ is called a *real vector space;* similarly if $\mathcal{F}$ is $\mathcal{Q}$ or if $\mathcal{F}$ is $\mathcal{C}$, we speak of *rational vector spaces* or *complex vector spaces*.

§ 3. Examples

Before discussing the implications of the axioms, we give some examples. We shall refer to these examples over and over again, and we shall use the notation established here throughout the rest of our work.

(1) Let $\mathcal{C}^1(= \mathcal{C})$ be the set of all complex numbers; if we interpret $x + y$ and αx as ordinary complex numerical addition and multiplication, $\mathcal{C}^1$ becomes a complex vector space.

(2) Let $\mathcal{P}$ be the set of all polynomials, with complex coefficients, in a variable t. To make $\mathcal{P}$ into a complex vector space, we interpret vector addition and scalar multiplication as the ordinary addition of two polynomials and the multiplication of a polynomial by a complex number; the origin in $\mathcal{P}$ is the polynomial identically zero.

Example (1) is too simple and example (2) is too complicated to be typical of the main contents of this book. We give now another example of complex vector spaces which (as we shall see later) is general enough for all our purposes.

(3) Let $\mathcal{C}^n$, $n = 1, 2, \cdots$, be the set of all n-tuples of complex numbers. If $x = (\xi_1, \cdots, \xi_n)$ and $y = (\eta_1, \cdots, \eta_n)$ are elements of $\mathcal{C}^n$, we write, by definition,

$$x + y = (\xi_1 + \eta_1, \cdots, \xi_n + \eta_n),$$

$$\alpha x = (\alpha\xi_1, \cdots, \alpha\xi_n),$$

$$0 = (0, \cdots, 0),$$

$$-x = (-\xi_1, \cdots, -\xi_n).$$

It is easy to verify that all parts of our axioms (A), (B), and (C), § 2, are satisfied, so that $\mathcal{C}^n$ is a complex vector space; it will be called *n-dimensional complex coordinate space*.

(4) For each positive integer n, let $\mathcal{P}_n$ be the set of all polynomials (with complex coefficients, as in example (2)) of degree $\leq n - 1$, together with the polynomial identically zero. (In the usual discussion of degree, the degree of this polynomial is not defined, so that we cannot say that it has degree $\leq n - 1$.) With the same interpretation of the linear operations (addition and scalar multiplication) as in (2), $\mathcal{P}_n$ is a complex vector space.

(5) A close relative of $\mathcal{C}^n$ is the set $\mathcal{R}^n$ of all n-tuples of real numbers. With the same formal definitions of addition and scalar multiplication as for $\mathcal{C}^n$, except that now we consider only real scalars α, the space $\mathcal{R}^n$ is a real vector space; it will be called *n-dimensional real coordinate space*.

(6) All the preceding examples can be generalized. Thus, for instance, an obvious generalization of (1) can be described by saying that every field may be regarded as a vector space over itself. A common generalization of (3) and (5) starts with an arbitrary field $\mathcal{F}$ and forms the set $\mathcal{F}^n$ of n-tuples of elements of $\mathcal{F}$; the formal definitions of the linear operations are the same as for the case $\mathcal{F} = \mathcal{C}$.

(7) A field, by definition, has at least two elements; a vector space, however, may have only one. Since every vector space contains an origin, there is essentially (i.e., except for notation) only one vector space having only one vector. This most trivial vector space will be denoted by $\mathcal{O}$.

(8) If, in the set $\mathcal{R}$ of all real numbers, addition is defined as usual and multiplication of a real number by a rational number is defined as usual, then $\mathcal{R}$ becomes a rational vector space.

(9) If, in the set $\mathcal{C}$ of all complex numbers, addition is defined as usual and multiplication of a complex number by a real number is defined as usual, then $\mathcal{C}$ becomes a real vector space. (Compare this example with (1); they are quite different.)

§ 4. Comments

A few comments are in order on our axioms and notation. There are striking similarities (and equally striking differences) between the axioms for a field and the axioms for a vector space over a field. In both cases, the axioms (A) describe the additive structure of the system, the axioms (B) describe its multiplicative structure, and the axioms (C) describe the connection between the two structures. Those familiar with algebraic terminology will have recognized the axioms (A) (in both § 1 and § 2) as the defining conditions of an abelian (commutative) group; the axioms (B) and (C) (in § 2) express the fact that the group admits scalars as operators. We mention in passing that if the scalars are elements of a ring (instead of a field), the generalized concept corresponding to a vector space is called a *module*.

Special real vector spaces (such as $\Re^2$ and $\Re^3$) are familiar in geometry. There seems at this stage to be no excuse for our apparently uninteresting insistence on fields other than $\Re$, and, in particular, on the field $\mathbb{C}$ of complex numbers. We hope that the reader is willing to take it on faith that we shall have to make use of deep properties of complex numbers later (conjugation, algebraic closure), and that in both the applications of vector spaces to modern (quantum mechanical) physics and the mathematical generalization of our results to Hilbert space, complex numbers play an important role. Their one great disadvantage is the difficulty of drawing pictures; the ordinary picture (Argand diagram) of $\mathbb{C}^1$ is indistinguishable from that of $\Re^2$, and a graphic representation of $\mathbb{C}^2$ seems to be out of human reach. On the occasions when we have to use pictorial language we shall therefore use the terminology of $\Re^n$ in $\mathbb{C}^n$, and speak of $\mathbb{C}^2$, for example, as a plane.

Finally we comment on notation. We observe that the symbol 0 has been used in two meanings: once as a scalar and once as a vector. To make the situation worse, we shall later, when we introduce linear functionals and linear transformations, give it still other meanings. Fortunately the relations among the various interpretations of 0 are such that, after this word of warning, no confusion should arise from this practice.

EXERCISES

1. Prove that if x and y are vectors and if α is a scalar, then the following relations hold.
    (a) $0 + x = x$.
    (b) $-0 = 0$.
    (c) $\alpha \cdot 0 = 0$.
    (d) $0 \cdot x = 0$. (Observe that the same symbol is used on both sides of this equation; on the left it denotes a scalar, on the right it denotes a vector.)
    (e) If $\alpha x = 0$, then either $\alpha = 0$ or $x = 0$ (or both).
    (f) $-x = (-1)x$.
    (g) $y + (x - y) = x$. (Here $x - y = x + (-y)$.)

2. If p is a prime, then $Z_p{}^n$ is a vector space over Z_p (cf. § 1, Ex. 3); how many vectors are there in this vector space?

3. Let $\mathcal{U}$ be the set of all (ordered) pairs of real numbers. If $x = (\xi_1, \xi_2)$ and $y = (\eta_1, \eta_2)$ are elements of $\mathcal{U}$, write

$$x + y = (\xi_1 + \eta_1, \xi_2 + \eta_2)$$

$$\alpha x = (\alpha \xi_1, 0)$$

$$0 = (0, 0)$$

$$-x = (-\xi_1, -\xi_2).$$

Is $\mathcal{U}$ a vector space with respect to these definitions of the linear operations? Why?

4. Sometimes a subset of a vector space is itself a vector space (with respect to the linear operations already given). Consider, for example, the vector space $\mathcal{C}^3$ and the subsets $\mathcal{U}$ of $\mathcal{C}^3$ consisting of those vectors (ξ_1, ξ_2, ξ_3) for which

(a) ξ_1 is real,
(b) $\xi_1 = 0$,
(c) either $\xi_1 = 0$ or $\xi_2 = 0$,
(d) $\xi_1 + \xi_2 = 0$,
(e) $\xi_1 + \xi_2 = 1$.

In which of these cases is $\mathcal{U}$ a vector space?

5. Consider the vector space $\mathcal{P}$ and the subsets $\mathcal{U}$ of $\mathcal{P}$ consisting of those vectors (polynomials) x for which

(a) x has degree 3,
(b) $2x(0) = x(1)$,
(c) $x(t) \geqq 0$ whenever $0 \leqq t \leqq 1$,
(d) $x(t) = x(1 - t)$ for all t.

In which of these cases is $\mathcal{U}$ a vector space?

§ 5. Linear dependence

Now that we have described the spaces we shall work with, we must specify the relations among the elements of those spaces that will be of interest to us.

We begin with a few words about the summation notation. If corresponding to each of a set of indices i there is given a vector x_i, and if it is not necessary or not convenient to specify the set of indices exactly, we shall simply speak of a set $\{x_i\}$ of vectors. (We admit the possibility that the same vector corresponds to two distinct indices. In all honesty, therefore, it should be stated that what is important is not which vectors appear in $\{x_i\}$, but how they appear.) If the index-set under consideration is finite, we shall denote the sum of the corresponding vectors by $\sum_i x_i$ (or, when desirable, by a more explicit symbol such as $\sum_{i=1}^n x_i$). In order to avoid frequent and fussy case distinctions, it is a good idea to admit into the general theory sums such as $\sum_i x_i$ even when there are no indices i to be summed over, or, more precisely, even when the index-set under consideration is empty. (In that case, of course, there are no vectors to sum, or, more precisely, the set $\{x_i\}$ is also empty.) The value of such an "empty sum" is defined, naturally enough, to be the vector 0.

DEFINITION. A finite set $\{x_i\}$ of vectors is *linearly dependent* if there exists a corresponding set $\{\alpha_i\}$ of scalars, not all zero, such that

$$\sum_i \alpha_i x_i = 0.$$

If, on the other hand, $\sum_i \alpha_i x_i = 0$ implies that $\alpha_i = 0$ for each i, the set $\{x_i\}$ is *linearly independent*.

The wording of this definition is intended to cover the case of the empty set; the result in that case, though possibly paradoxical, dovetails very satisfactorily with the rest of the theory. The result is that the empty set of vectors is linearly independent. Indeed, if there are no indices i, then it is not possible to pick out some of them and to assign to the selected ones a non-zero scalar so as to make a certain sum vanish. The trouble is not in avoiding the assignment of zero; it is in finding an index to which something can be assigned. Note that this argument shows that the empty set is not linearly dependent; for the reader not acquainted with arguing by "vacuous implication," the equivalence of the definition of linear independence with the straightforward negation of the definition of linear dependence needs a little additional intuitive justification. The easiest way to feel comfortable about the assertion "$\sum_i \alpha_i x_i = 0$ implies that $\alpha_i = 0$ for each i," in case there are no indices i, is to rephrase it this way: "if $\sum_i \alpha_i x_i = 0$, then there is no index i for which $\alpha_i \neq 0$." This version is obviously true if there is no index i at all.

Linear dependence and independence are properties of sets of vectors; it is customary, however, to apply the adjectives to vectors themselves, and thus we shall sometimes say "a set of linearly independent vectors" instead of "a linearly independent set of vectors." It will be convenient also to speak of the linear dependence and independence of a not necessarily finite set, $\mathfrak{X}$, of vectors. We shall say that $\mathfrak{X}$ is linearly independent if every finite subset of $\mathfrak{X}$ is such; otherwise $\mathfrak{X}$ is linearly dependent.

To gain insight into the meaning of linear dependence, let us study the examples of vector spaces that we already have.

(1) If x and y are any two vectors in $\mathbb{C}^1$, then x and y form a linearly dependent set. If $x = y = 0$, this is trivial; if not, then we have, for example, the relation $yx + (-x)y = 0$. Since it is clear that every set containing a linearly dependent subset is itself linearly dependent, this shows that in $\mathbb{C}^1$ every set containing more than one element is a linearly dependent set.

(2) More interesting is the situation in the space $\mathcal{P}$. The vectors x, y, and z, defined by

$$x(t) = 1 - t,$$

$$y(t) = t(1 - t),$$

$$z(t) = 1 - t^2,$$

are, for example, linearly dependent, since $x + y - z = 0$. However, the infinite set of vectors x_0, x_1, x_2, $\cdots$, defined by

$$x_0(t) = 1, \quad x_1(t) = t, \quad x_2(t) = t^2, \cdots,$$

is a linearly independent set, for if we had any relation of the form

$$\alpha_0 x_0 + \alpha_1 x_1 + \cdots + \alpha_n x_n = 0,$$

then we should have a polynomial identity

$$\alpha_0 + \alpha_1 t + \cdots + \alpha_n t^n = 0,$$

whence $\qquad\qquad \alpha_0 = \alpha_1 = \cdots = \alpha_n = 0.$

(3) As we mentioned before, the spaces $\mathfrak{C}^n$ are the prototype of what we want to study; let us examine, for example, the case $n = 3$. To those familiar with higher-dimensional geometry, the notion of linear dependence in this space (or, more properly speaking, in its real analogue $\mathfrak{R}^3$) has a concrete geometric meaning, which we shall only mention. In geometrical language, two vectors are linearly dependent if and only if they are collinear with the origin, and three vectors are linearly dependent if and only if they are coplanar with the origin. (If one thinks of a vector not as a point in a space but as an arrow pointing from the origin to some given point, the preceding sentence should be modified by crossing out the phrase "with the origin" both times that it occurs.) We shall presently introduce the notion of linear manifolds (or vector subspaces) in a vector space, and, in that connection, we shall occasionally use the language suggested by such geometrical considerations.

§ 6. Linear combinations

We shall say, whenever $x = \sum_i \alpha_i x_i$, that x is a *linear combination* of $\{x_i\}$; we shall use without any further explanation all the simple grammatical implications of this terminology. Thus we shall say, in case x is a linear combination of $\{x_i\}$, that x is linearly dependent on $\{x_i\}$; we shall leave to the reader the proof that if $\{x_i\}$ is linearly independent, then a necessary and sufficient condition that x be a linear combination of $\{x_i\}$ is that the enlarged set, obtained by adjoining x to $\{x_i\}$, be linearly dependent. Note that, in accordance with the definition of an empty sum, the origin is a linear combination of the empty set of vectors; it is, moreover, the only vector with this property.

The following theorem is the fundamental result concerning linear dependence.

THEOREM. *The set of non-zero vectors $x_1, \cdots, x_n$ is linearly dependent if and only if some x_k, $2 \leq k \leq n$, is a linear combination of the preceding ones.*

PROOF. Let us suppose that the vectors $x_1, \cdots, x_n$ are linearly dependent, and let k be the first integer between 2 and n for which $x_1, \cdots, x_k$ are linearly dependent. (If worse comes to worst, our assumption assures us that $k = n$ will do.) Then

$$\alpha_1 x_1 + \cdots + \alpha_k x_k = 0$$

for a suitable set of α's (not all zero); moreover, whatever the α's, we cannot have $\alpha_k = 0$, for then we should have a linear dependence relation among $x_1, \cdots, x_{k-1}$, contrary to the definition of k. Hence

$$x_k = \frac{-\alpha_1}{\alpha_k} x_1 + \cdots + \frac{-\alpha_{k-1}}{\alpha_k} x_{k-1},$$

as was to be proved. This proves the necessity of our condition; sufficiency is clear since, as we remarked before, every set containing a linearly dependent set is itself such.

§ 7. Bases

DEFINITION. A (linear) *basis* (or a *coordinate system*) in a vector space $\mathcal{U}$ is a set $\mathfrak{X}$ of linearly independent vectors such that every vector in $\mathcal{U}$ is a linear combination of elements of $\mathfrak{X}$. A vector space $\mathcal{U}$ is *finite-dimensional* if it has a finite basis.

Except for the occasional consideration of examples we shall restrict our attention, throughout this book, to finite-dimensional vector spaces.

For examples of bases we turn again to the spaces $\mathcal{O}$ and $\mathcal{C}^n$. In $\mathcal{O}$, the set $\{x_n\}$, where $x_n(t) = t^n$, $n = 0, 1, 2, \cdots$, is a basis; every polynomial is, by definition, a linear combination of a finite number of x_n. Moreover $\mathcal{O}$ has no finite basis, for, given any finite set of polynomials, we can find a polynomial of higher degree than any of them; this latter polynomial is obviously not a linear combination of the former ones.

An example of a basis in $\mathcal{C}^n$ is the set of vectors x_i, $i = 1, \cdots, n$, defined by the condition that the j-th coordinate of x_i is δ_{ij}. (Here we use for the first time the popular Kronecker δ; it is defined by $\delta_{ij} = 1$ if $i = j$ and $\delta_{ij} = 0$ if $i \neq j$.) Thus we assert that in $\mathcal{C}^3$ the vectors $x_1 = (1, 0, 0)$, $x_2 = (0, 1, 0)$, and $x_3 = (0, 0, 1)$ form a basis. It is easy to see that they are linearly independent; the formula

$$x = (\xi_1, \xi_2, \xi_3) = \xi_1 x_1 + \xi_2 x_2 + \xi_3 x_3$$

proves that every x in $\mathcal{C}^3$ is a linear combination of them.

In a general finite-dimensional vector space $\mathcal{U}$, with basis $\{x_1, \cdots, x_n\}$, we know that every x can be written in the form

$$x = \sum_i \xi_i x_i;$$

we assert that the ξ's are uniquely determined by x. The proof of this assertion is an argument often used in the theory of linear dependence. If we had $x = \sum_i \eta_i x_i$, then we should have, by subtraction,

$$\sum_i (\xi_i - \eta_i)x_i = 0.$$

Since the x_i are linearly independent, this implies that $\xi_i - \eta_i = 0$ for $i = 1, \cdots, n$; in other words, the ξ's are the same as the η's. (Observe that writing $\{x_1, \cdots, x_n\}$ for a basis with n elements is not the proper thing to do in case $n = 0$. We shall, nevertheless, frequently use this notation. Whenever that is done, it is, in principle, necessary to adjoin a separate discussion designed to cover the vector space Θ. In fact, however, everything about that space is so trivial that the details are not worth writing down, and we shall omit them.)

THEOREM. *If $\mathcal{V}$ is a finite-dimensional vector space and if $\{y_1, \cdots, y_m\}$ is any set of linearly independent vectors in $\mathcal{V}$, then, unless the y's already form a basis, we can find vectors $y_{m+1}, \cdots, y_{m+p}$ so that the totality of the y's, that is, $\{y_1, \cdots, y_m, y_{m+1}, \cdots, y_{m+p}\}$, is a basis. In other words, every linearly independent set can be extended to a basis.*

PROOF. Since $\mathcal{V}$ is finite-dimensional, it has a finite basis, say $\{x_1, \cdots, x_n\}$. We consider the set $\mathcal{S}$ of vectors

$$y_1, \cdots, y_m, x_1, \cdots, x_n,$$

in this order, and we apply to this set the theorem of § 6 several times in succession. In the first place, the set $\mathcal{S}$ is linearly dependent, since the y's are (as are all vectors) linear combinations of the x's. Hence some vector of $\mathcal{S}$ is a linear combination of the preceding ones; let z be the first such vector. Then z is different from any y_i, $i = 1, \cdots, m$ (since the y's are linearly independent), so that z is equal to some x, say $z = x_i$. We consider the new set $\mathcal{S}'$ of vectors

$$y_1, \cdots, y_m, x_1, \cdots, x_{i-1}, x_{i+1}, \cdots, x_n.$$

We observe that every vector in $\mathcal{V}$ is a linear combination of vectors in $\mathcal{S}'$, since by means of $y_1, \cdots, y_m, x_1, \cdots, x_{i-1}$ we may express x_i, and then by means of $x_1, \cdots, x_{i-1}, x_i, x_{i+1}, \cdots, x_n$ we may express any vector. (The x's form a basis.) If $\mathcal{S}'$ is linearly independent, we are done. If it is not, we apply the theorem of § 6 again and again the same way till we reach a linearly independent set containing $y_1, \cdots, y_m$, in terms of which we may express every vector in $\mathcal{V}$. This last set is a basis containing the y's.

1. (a) Prove that the four vectors

$$x = (1, 0, 0),$$

$$y = (0, 1, 0),$$

$$z = (0, 0, 1),$$

$$u = (1, 1, 1),$$

in $\mathbb{C}^3$ form a linearly dependent set, but any three of them are linearly independent. ('To test the linear dependence of vectors $x = (\xi_1, \xi_2, \xi_3)$, $y = (\eta_1, \eta_2, \eta_3)$, and $z = (\zeta_1, \zeta_2, \zeta_3)$ in $\mathbb{C}^3$, proceed as follows. Assume that α, β, and γ can be found so that $\alpha x + \beta y + \gamma z = 0$. This means that

$$\alpha\xi_1 + \beta\eta_1 + \gamma\zeta_1 = 0,$$

$$\alpha\xi_2 + \beta\eta_2 + \gamma\zeta_2 = 0,$$

$$\alpha\xi_3 + \beta\eta_3 + \gamma\zeta_3 = 0.$$

The vectors x, y, and z are linearly dependent if and only if these equations have a solution other than $\alpha = \beta = \gamma = 0$.)

(b) If the vectors x, y, z, and u in $\mathcal{P}$ are defined by $x(t) = 1$, $y(t) = t$, $z(t) = t^2$, and $u(t) = 1 + t + t^2$, prove that x, y, z, and u are linearly dependent, but any three of them are linearly independent.

2. Prove that if $\mathcal{R}$ is considered as a rational vector space (see § 3, (8)), then a necessary and sufficient condition that the vectors 1 and ξ in $\mathcal{R}$ be linearly independent is that the real number ξ be irrational.

3. Is it true that if x, y, and z are linearly independent vectors, then so also are $x + y$, $y + z$, and $z + x$?

4. (a) Under what conditions on the scalar ξ are the vectors $(1 + \xi, 1 - \xi)$ and $(1 - \xi, 1 + \xi)$ in $\mathbb{C}^2$ linearly dependent?

(b) Under what conditions on the scalar ξ are the vectors $(\xi, 1, 0)$, $(1, \xi, 1)$, and $(0, 1, \xi)$ in $\mathcal{R}^3$ linearly dependent?

(c) What is the answer to (b) for $\mathbb{Q}^3$ (in place of $\mathcal{R}^3$)?

5. (a) The vectors (ξ_1, ξ_2) and (η_1, η_2) in $\mathbb{C}^2$ are linearly dependent if and only if $\xi_1\eta_2 = \xi_2\eta_1$.

(b) Find a similar necessary and sufficient condition for the linear dependence of two vectors in $\mathbb{C}^3$. Do the same for three vectors in $\mathbb{C}^3$.

(c) Is there a set of three linearly independent vectors in $\mathbb{C}^2$?

6. (a) Under what conditions on the scalars ξ and η are the vectors $(1, \xi)$ and $(1, \eta)$ in $\mathbb{C}^2$ linearly dependent?

(b) Under what conditions on the scalars ξ, η, and ζ are the vectors $(1, \xi, \xi^2)$, $(1, \eta, \eta^2)$, and $(1, \zeta, \zeta^2)$ in $\mathbb{C}^3$ linearly dependent?

(c) Guess and prove a generalization of (a) and (b) to $\mathbb{C}^n$.

7. (a) Find two bases in $\mathbb{C}^4$ such that the only vectors common to both are $(0, 0, 1, 1)$ and $(1, 1, 0, 0)$.

(b) Find two bases in $\mathbb{C}^4$ that have no vectors in common so that one of them contains the vectors $(1, 0, 0, 0)$ and $(1, 1, 0, 0)$ and the other one contains the vectors $(1, 1, 1, 0)$ and $(1, 1, 1, 1)$.

8. (a) Under what conditions on the scalar ξ do the vectors $(1, 1, 1)$ and $(1, \xi, \xi^2)$ form a basis of $\mathbb{C}^3$?

(b) Under what conditions on the scalar ξ do the vectors $(0, 1, \xi)$, $(\xi, 0, 1)$, and $(\xi, 1, 1 + \xi)$ form a basis of $\mathbb{C}^3$?

9. Consider the set of all those vectors in $\mathbb{C}^3$ each of whose coordinates is either 0 or 1; how many different bases does this set contain?

10. If $\mathfrak{X}$ is the set consisting of the six vectors $(1, 1, 0, 0)$, $(1, 0, 1, 0)$, $(1, 0, 0, 1)$, $(0, 1, 1, 0)$, $(0, 1, 0, 1)$, $(0, 0, 1, 1)$ in $\mathbb{C}^4$, find two different maximal linearly independent subsets of $\mathfrak{X}$. (A maximal linearly independent subset of $\mathfrak{X}$ is a linearly independent subset $\mathfrak{Y}$ of $\mathfrak{X}$ that becomes linearly dependent every time that a vector of $\mathfrak{X}$ that is not already in $\mathfrak{Y}$ is adjoined to $\mathfrak{Y}$.)

11. Prove that every vector space has a basis. (The proof of this fact is out of reach for those not acquainted with some transfinite trickery, such as well-ordering or Zorn's lemma.)

§ 8. Dimension

THEOREM 1. *The number of elements in any basis of a finite-dimensional vector space $\mathcal{V}$ is the same as in any other basis.*

PROOF. The proof of this theorem is a slight refinement of the method used in § 6, and, incidentally, it proves something more than the theorem states. Let $\mathfrak{X} = \{x_1, \cdots, x_n\}$ and $\mathfrak{Y} = \{y_1, \cdots, y_m\}$ be two finite sets of vectors, each with one of the two defining properties of a basis; i.e., we assume that every vector in $\mathcal{V}$ is a linear combination of the x's (but not that the x's are linearly independent), and we assume that the y's are linearly independent (but not that every vector is a linear combination of them). We may apply the theorem of § 6, just as above, to the set $\mathcal{S}$ of vectors

$$y_m, x_1, \cdots, x_n.$$

Again we know that every vector is a linear combination of vectors of $\mathcal{S}$ and that $\mathcal{S}$ is linearly dependent. Reasoning just as before, we obtain a set $\mathcal{S}'$ of vectors

$$y_m, x_1, \cdots, x_{i-1}, x_{i+1}, \cdots, x_n,$$

again with the property that every vector is a linear combination of vectors of $\mathcal{S}'$. Now we write y_{m-1} in front of the vectors of $\mathcal{S}'$ and apply the same argument. Continuing in this way, we see that the x's will not be exhausted before the y's, since otherwise the remaining y's would have to be linear combinations of the ones already incorporated into $\mathcal{S}$, whereas we know

that the y's are linearly independent. In other words, after the argument
has been applied m times, we obtain a set with the same property the
x's had, and this set differs from the set of x's in that m of them are replaced by y's. This seemingly innocent statement is what we are after;
it implies that $n \geqq m$. Consequently if both $\mathfrak{X}$ and $\mathfrak{Y}$ are bases (so that
they each have both properties), then $n \geqq m$ and $m \geqq n$.

DEFINITION. The *dimension* of a finite-dimensional vector space $\mathfrak{V}$ is
the number of elements in a basis of $\mathfrak{V}$.

Observe that since the empty set of vectors is a basis of the trivial
space $\mathfrak{O}$, the definition implies that that space has dimension 0. At the
same time the definition (together with the fact that we have already
exhibited, in § 7, one particular basis of $\mathfrak{C}^n$) at last justifies our terminology
and enables us to announce the pleasant result: n-dimensional coordinate
space is n-dimensional. (Since the argument is the same for $\mathfrak{R}^n$ and for
$\mathfrak{C}^n$, the assertion is true in both the real case and the complex case.)
Our next result is a corollary of Theorem 1 (via the theorem of § 7).

THEOREM 2. *Every set of $n + 1$ vectors in an n-dimensional vector space
$\mathfrak{V}$ is linearly dependent. A set of n vectors in $\mathfrak{V}$ is a basis if and only if it is
linearly independent, or, alternatively, if and only if every vector in $\mathfrak{V}$
is a linear combination of elements of the set.*

§ 9. Isomorphism

As an application of the notion of linear basis, or coordinate system,
we shall now fulfill an implicit earlier promise by showing that every
finite-dimensional vector space over a field $\mathfrak{F}$ is essentially the same as
(in technical language, is isomorphic to) some $\mathfrak{F}^n$.

DEFINITION. Two vector spaces $\mathfrak{U}$ and $\mathfrak{V}$ (over the same field) are
isomorphic if there is a one-to-one correspondence between the vectors
x of $\mathfrak{U}$ and the vectors y of $\mathfrak{V}$, say $y = T(x)$, such that

$$T(\alpha_1 x_1 + \alpha_2 x_2) = \alpha_1 T(x_1) + \alpha_2 T(x_2).$$

In other words, $\mathfrak{U}$ and $\mathfrak{V}$ are isomorphic if there is an isomorphism (such
as T) between them, where an *isomorphism* is a one-to-one correspondence
that preserves all linear relations.

It is easy to see that isomorphic finite-dimensional vector spaces have
the same dimension; to each basis in one space there corresponds a basis
in the other space. Thus dimension is an isomorphism invariant; we shall
now show that it is the only isomorphism invariant, in the sense that every

two vector spaces with the same finite dimension (over the same field, of course) are isomorphic. Since the isomorphism of $\mathcal{U}$ and $\mathcal{V}$ on the one hand, and of $\mathcal{V}$ and $\mathcal{W}$ on the other hand, implies that $\mathcal{U}$ and $\mathcal{W}$ are isomorphic, it will be sufficient to prove the following theorem.

THEOREM. *Every n-dimensional vector space* $\mathcal{V}$ *over a field* $\mathfrak{F}$ *is isomorphic to* $\mathfrak{F}^n$.

PROOF. Let $\{x_1, \cdots, x_n\}$ be any basis in $\mathcal{V}$. Each x in $\mathcal{V}$ can be written in the form $\xi_1 x_1 + \cdots + \xi_n x_n$, and we know that the scalars $\xi_1, \cdots, \xi_n$ are uniquely determined by x. We consider the one-to-one correspondence

$$x \rightleftarrows (\xi_1, \cdots, \xi_n)$$

between $\mathcal{V}$ and $\mathfrak{F}^n$. If $y = \eta_1 x_1 + \cdots + \eta_n x_n$, then

$$\alpha x + \beta y = (\alpha \xi_1 + \beta \eta_1) x_1 + \cdots + (\alpha \xi_n + \beta \eta_n) x_n;$$

this establishes the desired isomorphism.

One might be tempted to say that from now on it would be silly to try to preserve an appearance of generality by talking of the general n-dimensional vector space, since we know that, from the point of view of studying linear problems, isomorphic vector spaces are indistinguishable, and, consequently, we might as well always study $\mathfrak{F}^n$. There is one catch. The most important properties of vectors and vector spaces are the ones that are independent of coordinate systems, or, in other words, the ones that are invariant under isomorphisms. The correspondence between $\mathcal{V}$ and $\mathfrak{F}^n$ was, however, established by choosing a coordinate system; were we always to study $\mathfrak{F}^n$, we would always be tied down to that particular coordinate system, or else we would always be faced with the chore of showing that our definitions and theorems are independent of the coordinate system in which they happen to be stated. (This horrible dilemma will become clear later, on the few occasions when we shall be forced to use a particular coordinate system to give a definition.) Accordingly, in the greater part of this book, we shall ignore the theorem just proved, and we shall treat n-dimensional vector spaces as self-respecting entities, independently of any basis. Besides the reasons just mentioned, there is another reason for doing this: many special examples of vector spaces, such for instance as $\mathcal{P}_n$, would lose a lot of their intuitive content if we were to transform them into $\mathcal{C}^n$ and speak of coordinates only. In studying vector spaces, such as $\mathcal{P}_n$, and their relation to other vector spaces, we must be able to handle them with equal ease in different coordinate systems, or, and this is essentially the same thing, we must be able to handle them without using any coordinate systems at all.

1. (a) What is the dimension of the set $\mathcal{C}$ of all complex numbers considered as a real vector space? (See § 3, (9).)

(b) Every complex vector space $\mathcal{U}$ is intimately associated with a real vector space $\mathcal{U}^-$; the space $\mathcal{U}^-$ is obtained from $\mathcal{U}$ by refusing to multiply vectors of $\mathcal{U}$ by anything other than real scalars. If the dimension of the complex vector space $\mathcal{U}$ is n, what is the dimension of the real vector space $\mathcal{U}^-$?

2. Is the set $\mathcal{R}$ of all real numbers a finite-dimensional vector space over the field $\mathcal{Q}$ of all rational numbers? (See § 3, (8). The question is not trivial; it helps to know something about cardinal numbers.)

3. How many vectors are there in an n-dimensional vector space over the field Z_p (where p is a prime)?

4. Discuss the following assertion: if two rational vector spaces have the same cardinal number (i.e., if there is some one-to-one correspondence between them), then they are isomorphic (i.e., there is a linearity-preserving one-to-one correspondence between them). A knowledge of the basic facts of cardinal arithmetic is needed for an intelligent discussion.

§ 10. Subspaces

The objects of interest in geometry are not only the points of the space under consideration, but also its lines, planes, etc. We proceed to study the analogues, in general vector spaces, of these higher-dimensional elements.

DEFINITION. A non-empty subset $\mathfrak{M}$ of a vector space $\mathcal{U}$ is a *subspace* or a *linear manifold* if along with every pair, x and y, of vectors contained in $\mathfrak{M}$, every linear combination $\alpha x + \beta y$ is also contained in $\mathfrak{M}$.

A word of warning: along with each vector x, a subspace also contains $x - x$. Hence if we interpret subspaces as generalized lines and planes, we must be careful to consider only lines and planes that pass through the origin.

A subspace $\mathfrak{M}$ in a vector space $\mathcal{U}$ is itself a vector space; the reader can easily verify that, with the same definitions of addition and scalar multiplication as we had in $\mathcal{U}$, the set satisfies the axioms (A), (B), and (C) of § 2.

Two special examples of subspaces are: (i) the set $\mathcal{O}$ consisting of the origin only, and (ii) the whole space $\mathcal{U}$. The following examples are less trivial.

(1) Let n and m be any two strictly positive integers, $m \leq n$. Let $\mathfrak{M}$ be the set of all vectors $x = (\xi_1, \cdots, \xi_n)$ in $\mathcal{C}^n$ for which $\xi_1 = \cdots = \xi_m = 0$.

(2) With m and n as in (1), we consider the space $\mathcal{P}_n$, and any m real numbers $t_1, \cdots, t_m$. Let $\mathfrak{M}$ be the set of all vectors (polynomials) x in $\mathcal{P}_n$ for which $x(t_1) = \cdots = x(t_m) = 0$.

(3) Let $\mathfrak{M}$ be the set of all vectors x in $\mathcal{P}$ for which $x(t) = x(-t)$ holds identically in t.

We need some notation and some terminology. For any collection $\{\mathfrak{M}_\nu\}$ of subsets of a given set (say, for example, for a collection of subspaces in a vector space $\mathcal{V}$), we write $\bigcap_\nu \mathfrak{M}_\nu$ for the *intersection* of all $\mathfrak{M}_\nu$, i.e., for the set of points common to them all. Also, if $\mathfrak{M}$ and $\mathfrak{N}$ are subsets of a set, we write $\mathfrak{M} \subset \mathfrak{N}$ if $\mathfrak{M}$ is a subset of $\mathfrak{N}$, that is, if every element of $\mathfrak{M}$ lies in $\mathfrak{N}$ also. (Observe that we do not exclude the possibility $\mathfrak{M} = \mathfrak{N}$; thus we write $\mathcal{V} \subset \mathcal{V}$ as well as $\mathcal{O} \subset \mathcal{V}$.) For a finite collection $\{\mathfrak{M}_1, \cdots, \mathfrak{M}_n\}$, we shall write $\mathfrak{M}_1 \cap \cdots \cap \mathfrak{M}_n$ in place of $\bigcap_\nu \mathfrak{M}_\nu$; in case two subspaces $\mathfrak{M}$ and $\mathfrak{N}$ are such that $\mathfrak{M} \cap \mathfrak{N} = \mathcal{O}$, we shall say that $\mathfrak{M}$ and $\mathfrak{N}$ are *disjoint*.

§ 11. Calculus of subspaces

Theorem 1. *The intersection of any collection of subspaces is a subspace.*

proof. If we use an index ν to tell apart the members of the collection, so that the given subspaces are $\mathfrak{M}_\nu$, let us write

$$\mathfrak{M} = \bigcap_\nu \mathfrak{M}_\nu.$$

Since every $\mathfrak{M}_\nu$ contains 0, so does $\mathfrak{M}$, and therefore $\mathfrak{M}$ is not empty. If x and y belong to $\mathfrak{M}$ (that is, to all $\mathfrak{M}_\nu$), then $\alpha x + \beta y$ belongs to all $\mathfrak{M}_\nu$, and therefore $\mathfrak{M}$ is a subspace.

To see an application of this theorem, suppose that $\mathcal{S}$ is an arbitrary set of vectors (not necessarily a subspace) in a vector space $\mathcal{V}$. There certainly exist subspaces $\mathfrak{M}$ containing every element of $\mathcal{S}$ (that is, such that $\mathcal{S} \subset \mathfrak{M}$); the whole space $\mathcal{V}$ is, for example, such a subspace. Let $\mathfrak{M}$ be the intersection of all the subspaces containing $\mathcal{S}$; it is clear that $\mathfrak{M}$ itself is a subspace containing $\mathcal{S}$. It is clear, moreover, that $\mathfrak{M}$ is the smallest such subspace; if $\mathcal{S}$ is also contained in the subspace $\mathfrak{N}$, $\mathcal{S} \subset \mathfrak{N}$, then $\mathfrak{M} \subset \mathfrak{N}$. The subspace $\mathfrak{M}$ so defined is called the subspace *spanned* by $\mathcal{S}$ or the *span* of $\mathcal{S}$. The following result establishes the connection between the notion of spanning and the concepts studied in §§ 5–9.

Theorem 2. *If $\mathcal{S}$ is any set of vectors in a vector space $\mathcal{V}$ and if $\mathfrak{M}$ is the subspace spanned by $\mathcal{S}$, then $\mathfrak{M}$ is the same as the set of all linear combinations of elements of $\mathcal{S}$.*

PROOF. It is clear that a linear combination of linear combinations of elements of S may again be written as a linear combination of elements of S. Hence the set of all linear combinations of elements of S is a subspace containing S; it follows that this subspace must also contain $\mathfrak{M}$. Now turn the argument around: $\mathfrak{M}$ contains S and is a subspace; hence $\mathfrak{M}$ contains all linear combinations of elements of S.

We see therefore that in our new terminology we may define a linear basis as a set of linearly independent vectors that spans the whole space.

Our next result is an easy consequence of Theorem 2; its proof may be safely left to the reader.

THEOREM 3. *If $\mathfrak{K}$ and $\mathfrak{K}$ are any two subspaces and if $\mathfrak{M}$ is the subspace spanned by $\mathfrak{K}$ and $\mathfrak{K}$ together, then $\mathfrak{M}$ is the same as the set of all vectors of the form $x + y$, with x in $\mathfrak{K}$ and y in $\mathfrak{K}$.*

Prompted by this theorem, we shall use the notation $\mathfrak{K} + \mathfrak{K}$ for the subspace $\mathfrak{M}$ spanned by $\mathfrak{K}$ and $\mathfrak{K}$. We shall say that a subspace $\mathfrak{K}$ of a vector space $\mathcal{V}$ is a *complement* of a subspace $\mathfrak{K}$ if $\mathfrak{K} \cap \mathfrak{K} = \Theta$ and $\mathfrak{K} + \mathfrak{K} = \mathcal{V}$.

§ 12. Dimension of a subspace

THEOREM 1. *A subspace $\mathfrak{M}$ in an n-dimensional vector space $\mathcal{V}$ is a vector space of dimension $\leq n$.*

PROOF. It is possible to give a deceptively short proof of this theorem that runs as follows. Every set of $n + 1$ vectors in $\mathcal{V}$ is linearly dependent, hence the same is true of $\mathfrak{M}$; hence, in particular, the number of elements in each basis of $\mathfrak{M}$ is $\leq n$, Q.E.D.

The trouble with this argument is that we defined dimension n by requiring in the first place that there exist a finite basis, and then demanding that this basis contain exactly n elements. The proof above shows only that no basis can contain more than n elements; it does not show that any basis exists. Once the difficulty is observed, however, it is easy to fill the gap. If $\mathfrak{M} = \Theta$, then $\mathfrak{M}$ is 0-dimensional, and we are done. If $\mathfrak{M}$ contains a non-zero vector x_1, let $\mathfrak{M}_1$ ($\subset \mathfrak{M}$) be the subspace spanned by x_1. If $\mathfrak{M} = \mathfrak{M}_1$, then $\mathfrak{M}$ is 1-dimensional, and we are done. If $\mathfrak{M} \neq \mathfrak{M}_1$, let x_2 be an element of $\mathfrak{M}$ not contained in $\mathfrak{M}_1$, and let $\mathfrak{M}_2$ be the subspace spanned by x_1 and x_2; and so on. Now we may legitimately employ the argument given above; after no more than n steps of this sort, the process reaches an end, since (by § 8, Theorem 2) we cannot find $n + 1$ linearly independent vectors.

The following result is an important consequence of this second and correct proof of Theorem 1.

THEOREM 2. *Given any m-dimensional subspace $\mathfrak{M}$ in an n-dimensional vector space $\mathcal{V}$, we can find a basis $\{x_1, \cdots, x_m, x_{m+1}, \cdots, x_n\}$ in $\mathcal{V}$ so that $x_1, \cdots, x_m$ are in $\mathfrak{M}$ and form, therefore, a basis of $\mathfrak{M}$.*

We shall denote the dimension of a vector space $\mathcal{V}$ by the symbol dim $\mathcal{V}$. In this notation Theorem 1 asserts that if $\mathfrak{M}$ is a subspace of a finite-dimensional vector space $\mathcal{V}$, then dim $\mathfrak{M} \leqq$ dim $\mathcal{V}$.

EXERCISES

1. If $\mathfrak{M}$ and $\mathfrak{N}$ are finite-dimensional subspaces with the same dimension, and if $\mathfrak{M} \subset \mathfrak{N}$, then $\mathfrak{M} = \mathfrak{N}$.

2. If $\mathfrak{M}$ and $\mathfrak{N}$ are subspaces of a vector space $\mathcal{V}$, and if every vector in $\mathcal{V}$ belongs either to $\mathfrak{M}$ or to $\mathfrak{N}$ (or both), then either $\mathfrak{M} = \mathcal{V}$ or $\mathfrak{N} = \mathcal{V}$ (or both).

3. If x, y, and z are vectors such that $x + y + z = 0$, then x and y span the same subspace as y and z.

4. Suppose that x and y are vectors and $\mathfrak{M}$ is a subspace in a vector space $\mathcal{V}$; let $\mathcal{K}$ be the subspace spanned by $\mathfrak{M}$ and x, and let $\mathcal{K}$ be the subspace spanned by $\mathfrak{M}$ and y. Prove that if y is in $\mathcal{K}$ but not in $\mathfrak{M}$, then x is in $\mathcal{K}$.

5. Suppose that $\mathcal{L}$, $\mathfrak{M}$, and $\mathfrak{N}$ are subspaces of a vector space.
(a) Show that the equation

$$\mathcal{L} \cap (\mathfrak{M} + \mathfrak{N}) = (\mathcal{L} \cap \mathfrak{M}) + (\mathcal{L} \cap \mathfrak{N})$$

is not necessarily true.
(b) Prove that

$$\mathcal{L} \cap (\mathfrak{M} + (\mathcal{L} \cap \mathfrak{N})) = (\mathcal{L} \cap \mathfrak{M}) + (\mathcal{L} \cap \mathfrak{N}).$$

6. (a) Can it happen that a non-trivial subspace of a vector space $\mathcal{V}$ (i.e., a subspace different from both Θ and $\mathcal{V}$) has a unique complement?
(b) If $\mathfrak{M}$ is an m-dimensional subspace in an n-dimensional vector space, then every complement of $\mathfrak{M}$ has dimension $n - m$.

7. (a) Show that if both $\mathfrak{M}$ and $\mathfrak{N}$ are three-dimensional subspaces of a five-dimensional vector space, then $\mathfrak{M}$ and $\mathfrak{N}$ are not disjoint.
(b) If $\mathfrak{M}$ and $\mathfrak{N}$ are finite-dimensional subspaces of a vector space, then

$$\dim \mathfrak{M} + \dim \mathfrak{N} = \dim (\mathfrak{M} + \mathfrak{N}) + \dim (\mathfrak{M} \cap \mathfrak{N}).$$

8. A polynomial x is called *even* if $x(-t) = x(t)$ identically in t (see § 10, (3)), and it is called *odd* if $x(-t) = -x(t)$.
(a) Both the class $\mathfrak{M}$ of even polynomials and the class $\mathfrak{N}$ of odd polynomials are subspaces of the space $\mathcal{P}$ of all (complex) polynomials.
(b) Prove that $\mathfrak{M}$ and $\mathfrak{N}$ are each other's complements.

§ 13. Dual spaces

DEFINITION. A *linear functional* on a vector space $\mathcal{V}$ is a scalar-valued function y defined for every vector x, with the property that (identically in the vectors x_1 and x_2 and the scalars α_1 and α_2)

$$y(\alpha_1 x_1 + \alpha_2 x_2) = \alpha_1 y(x_1) + \alpha_2 y(x_2).$$

Let us look at some examples of linear functionals.

(1) For $x = (\xi_1, \cdots, \xi_n)$ in $\mathbb{C}^n$, write $y(x) = \xi_1$. More generally, let $\alpha_1, \cdots, \alpha_n$ be any n scalars and write

$$y(x) = \alpha_1 \xi_1 + \cdots + \alpha_n \xi_n.$$

We observe that for any linear functional y on any vector space

$$y(0) = y(0 \cdot 0) = 0 \cdot y(0) = 0;$$

for this reason a linear functional, as we defined it, is sometimes called *homogeneous*. In particular in $\mathbb{C}^n$, if y is defined by

$$y(x) = \alpha_1 \xi_1 + \cdots + \alpha_n \xi_n + \beta,$$

then y is not a linear functional unless $\beta = 0$.

(2) For any polynomial x in $\mathcal{P}$, write $y(x) = x(0)$. More generally, let $\alpha_1, \cdots, \alpha_n$ be any n scalars, let $t_1, \cdots, t_n$ be any n real numbers, and write

$$y(x) = \alpha_1 x(t_1) + \cdots + \alpha_n x(t_n).$$

Another example, in a sense a limiting case of the one just given, is obtained as follows. Let (a, b) be any finite interval on the real t-axis, and let α be any complex-valued integrable function defined on (a, b); define y by

$$y(x) = \int_a^b \alpha(t) x(t) \, dt.$$

(3) On an arbitrary vector space $\mathcal{V}$, define y by writing

$$y(x) = 0$$

for every x in $\mathcal{V}$.

The last example is the first hint of a general situation. Let $\mathcal{V}$ be any vector space and let $\mathcal{V}'$ be the collection of all linear functionals on $\mathcal{V}$. Let us denote by 0 the linear functional defined in (3) (compare the comment at the end of § 4). If y_1 and y_2 are linear functionals on $\mathcal{V}$ and if α_1 and α_2 are scalars, let us write y for the function defined by

$$y(x) = \alpha_1 y_1(x) + \alpha_2 y_2(x).$$

It is easy to see that y is a linear functional; we denote it by $\alpha_1 y_1 + \alpha_2 y_2$. With these definitions of the linear concepts (zero, addition, scalar multiplication), the set $\mathbb{U}'$ forms a vector space, the *dual space* of $\mathbb{U}$.

§ 14. Brackets

Before studying linear functionals and dual spaces in more detail, we wish to introduce a notation that may appear weird at first sight but that will clarify many situations later on. Usually we denote a linear functional by a single letter such as y. Sometimes, however, it is necessary to use the function notation fully and to indicate somehow that if y is a linear functional on $\mathbb{U}$ and if x is a vector in $\mathbb{U}$, then $y(x)$ is a particular scalar. According to the notation we propose to adopt here, we shall not write y followed by x in parentheses, but, instead, we shall write x and y enclosed between square brackets and separated by a comma. Because of the unusual nature of this notation, we shall expend on it some further verbiage.

As we have just pointed out $[x, y]$ is a substitute for the ordinary function symbol $y(x)$; both these symbols denote the scalar we obtain if we take the value of the linear function y at the vector x. Let us take an analogous situation (concerned with functions that are, however, not linear). Let y be the real function of a real variable defined for each real number x by $y(x) = x^2$. The notation $[x, y]$ is a symbolic way of writing down the recipe for actual operations performed; it corresponds to the sentence [take a number, and square it].

Using this notation, we may sum up: to every vector space $\mathbb{U}$ we make correspond the dual space $\mathbb{U}'$ consisting of all linear functionals on $\mathbb{U}$; to every pair, x and y, where x is a vector in $\mathbb{U}$ and y is a linear functional in $\mathbb{U}'$, we make correspond the scalar $[x, y]$ defined to be the value of y at x. In terms of the symbol $[x, y]$ the defining property of a linear functional is

(1) $$[\alpha_1 x_1 + \alpha_2 x_2, y] = \alpha_1[x_1, y] + \alpha_2[x_2, y],$$

and the definition of the linear operations for linear functionals is

(2) $$[x, \alpha_1 y_1 + \alpha_2 y_2] = \alpha_1[x, y_1] + \alpha_2[x, y_2].$$

The two relations together are expressed by saying that $[x, y]$ is a *bilinear functional* of the vectors x in $\mathbb{U}$ and y in $\mathbb{U}'$.

1. Consider the set $\mathcal{C}$ of complex numbers as a real vector space (as in § 3, (9)). Suppose that for each $x = \xi_1 + i\xi_2$ in $\mathcal{C}$ (where ξ_1 and ξ_2 are real numbers and $i = \sqrt{-1}$) the function y is defined by
   (a) $y(x) = \xi_1$,
   (b) $y(x) = \xi_2$,
   (c) $y(x) = \xi_1^2$,
   (d) $y(x) = \xi_1 - i\xi_2$,
   (e) $y(x) = \sqrt{\xi_1^2 + \xi_2^2}$.   (The square root sign attached to a positive number always denotes the positive square root of that number.)
In which of these cases is y a linear functional?

2. Suppose that for each $x = (\xi_1, \xi_2, \xi_3)$ in $\mathcal{C}^3$ the function y is defined by
   (a) $y(x) = \xi_1 + \xi_2$,
   (b) $y(x) = \xi_1 - \xi_3^2$,
   (c) $y(x) = \xi_1 + 1$,
   (d) $y(x) = \xi_1 - 2\xi_2 + 3\xi_3$.
In which of these cases is y a linear functional?

3. Suppose that for each x in $\mathcal{P}$ the function y is defined by

   (a) $y(x) = \displaystyle\int_{-1}^{+2} x(t)\, dt$,

   (b) $y(x) = \displaystyle\int_{0}^{2} (x(t))^2\, dt$,

   (c) $y(x) = \displaystyle\int_{0}^{1} t^2 x(t)\, dt$,

   (d) $y(x) = \displaystyle\int_{0}^{1} x(t^2)\, dt$,

   (e) $y(x) = \dfrac{dx}{dt}$,

   (f) $y(x) = \dfrac{d^2 x}{dt^2}\Big|_{t=1}$.

In which of these cases is y a linear functional?

4. If $(\alpha_0, \alpha_1, \alpha_2, \cdots)$ is an arbitrary sequence of complex numbers, and if x is an element of $\mathcal{P}$, $x(t) = \sum_{i=0}^{n} \xi_i t^i$, write $y(x) = \sum_{i=0}^{n} \xi_i \alpha_i$. Prove that y is an element of $\mathcal{P}'$ and that every element of $\mathcal{P}'$ can be obtained in this manner by a suitable choice of the α's.

5. If y is a non-zero linear functional on a vector space $\mathcal{V}$, and if α is an arbitrary scalar, does there necessarily exist a vector x in $\mathcal{V}$ such that $[x, y] = \alpha$?

6. Prove that if y and z are linear functionals (on the same vector space) such that $[x, y] = 0$ whenever $[x, z] = 0$, then there exists a scalar α such that $y = \alpha z$. (Hint: if $[x_0, z] \neq 0$, write $\alpha = [x_0, y]/[x_0, z]$.)

§ 15. Dual bases

One more word before embarking on the proofs of the important theorems. The concept of dual space was defined without any reference to coordinate systems; a glance at the following proofs will show a superabundance of coordinate systems. We wish to point out that this phenomenon is inevitable; we shall be establishing results concerning dimension, and dimension is the one concept (so far) whose very definition is given in terms of a basis.

THEOREM 1. *If $\mathcal{V}$ is an n-dimensional vector space, if $\{x_1, \cdots, x_n\}$ is a basis in $\mathcal{V}$, and if $\{\alpha_1, \cdots, \alpha_n\}$ is any set of n scalars, then there is one and only one linear functional y on $\mathcal{V}$ such that $[x_i, y] = \alpha_i$ for $i = 1, \cdots, n$.*

PROOF. Every x in $\mathcal{V}$ may be written in the form $x = \xi_1 x_1 + \cdots + \xi_n x_n$ in one and only one way; if y is any linear functional, then

$$[x, y] = \xi_1[x_1, y] + \cdots + \xi_n[x_n, y].$$

From this relation the uniqueness of y is clear; if $[x_i, y] = \alpha_i$, then the value of $[x, y]$ is determined, for every x, by $[x, y] = \sum_i \xi_i \alpha_i$. The argument can also be turned around; if we define y by

$$[x, y] = \xi_1 \alpha_1 + \cdots + \xi_n \alpha_n,$$

then y is indeed a linear functional, and $[x_i, y] = \alpha_i$.

THEOREM 2. *If $\mathcal{V}$ is an n-dimensional vector space and if $\mathfrak{X} = \{x_1, \cdots, x_n\}$ is a basis in $\mathcal{V}$, then there is a uniquely determined basis $\mathfrak{X}'$ in $\mathcal{V}'$, $\mathfrak{X}' = \{y_1, \cdots, y_n\}$, with the property that $[x_i, y_j] = \delta_{ij}$. Consequently the dual space of an n-dimensional space is n-dimensional.*

The basis $\mathfrak{X}'$ is called the *dual basis* of $\mathfrak{X}$.

PROOF. It follows from Theorem 1 that, for each $j = 1, \cdots, n$, a unique y_j in $\mathcal{V}'$ can be found so that $[x_i, y_j] = \delta_{ij}$; we have only to prove that the set $\mathfrak{X}' = \{y_1, \cdots, y_n\}$ is a basis in $\mathcal{V}'$.

In the first place, $\mathfrak{X}'$ is a linearly independent set, for if we had $\alpha_1 y_1 + \cdots + \alpha_n y_n = 0$, in other words, if

$$[x, \alpha_1 y_1 + \cdots + \alpha_n y_n] = \alpha_1[x, y_1] + \cdots + \alpha_n[x, y_n] = 0$$

for all x, then we should have, for $x = x_i$,

$$0 = \sum_j \alpha_j[x_i, y_j] = \sum_j \alpha_j \delta_{ij} = \alpha_i.$$

In the second place, every y in $\mathcal{V}'$ is a linear combination of $y_1, \cdots, y_n$. To prove this, write $[x_i, y] = \alpha_i$; then, for $x = \sum_i \xi_i x_i$, we have

$$[x, y] = \xi_1 \alpha_1 + \cdots + \xi_n \alpha_n.$$

On the other hand

$$[x, y_j] = \sum_i \xi_i [x_i, y_j] = \xi_j,$$

so that, substituting in the preceding equation, we get

$$[x, y] = \alpha_1 [x, y_1] + \cdots + \alpha_n [x, y_n]$$

$$= [x, \alpha_1 y_1 + \cdots + \alpha_n y_n].$$

Consequently $y = \alpha_1 y_1 + \cdots + \alpha_n y_n$, and the proof of the theorem is complete.

We shall need also the following easy consequence of Theorem 2.

THEOREM 3. *If u and v are any two different vectors of the n-dimensional vector space $\mathcal{V}$, then there exists a linear functional y on $\mathcal{V}$ such that $[u, y] \neq [v, y]$; or, equivalently, to any non-zero vector x in $\mathcal{V}$ there corresponds a y in $\mathcal{V}'$ such that $[x, y] \neq 0$.*

PROOF. That the two statements in the theorem are indeed equivalent is seen by considering $x = u - v$. We shall, accordingly, prove the latter statement only.

Let $\mathfrak{X} = \{x_1, \cdots, x_n\}$ be any basis in $\mathcal{V}$, and let $\mathfrak{X}' = \{y_1, \cdots, y_n\}$ be the dual basis in $\mathcal{V}'$. If $x = \sum_i \xi_i x_i$, then (as above) $[x, y_j] = \xi_j$. Hence if $[x, y] = 0$ for all y, and, in particular, if $[x, y_j] = 0$ for $j = 1, \cdots, n$, then $x = 0$.

§ 16. Reflexivity

It is natural to think that if the dual space $\mathcal{V}'$ of a vector space $\mathcal{V}$, and the relations between a space and its dual, are of any interest at all for $\mathcal{V}$, then they are of just as much interest for $\mathcal{V}'$. In other words, we propose now to form the dual space $(\mathcal{V}')'$ of $\mathcal{V}'$; for simplicity of notation we shall denote it by $\mathcal{V}''$. The verbal description of an element of $\mathcal{V}''$ is clumsy: such an element is a linear functional of linear functionals. It is, however, at this point that the greatest advantage of the notation $[x, y]$ appears; by means of it, it is easy to discuss $\mathcal{V}$ and its relation to $\mathcal{V}''$.

If we consider the symbol $[x, y]$ for some fixed $y = y_0$, we obtain nothing new: $[x, y_0]$ is merely another way of writing the value $y_0(x)$ of the function y_0 at the vector x. If, however, we consider the symbol $[x, y]$ for some fixed $x = x_0$, then we observe that the function of the vectors in $\mathcal{V}'$, whose value at y is $[x_0, y]$, is a scalar-valued function that happens to be linear (see § 14, (2)); in other words, $[x_0, y]$ defines a linear functional on $\mathcal{V}'$, and, consequently, an element of $\mathcal{V}''$.

By this method we have exhibited *some* linear functionals on $\mathcal{V}'$; have we exhibited them all? For the finite-dimensional case the following theorem furnishes the affirmative answer.

THEOREM. *If $\mathcal{V}$ is a finite-dimensional vector space, then corresponding to every linear functional z_0 on $\mathcal{V}'$ there is a vector x_0 in $\mathcal{V}$ such that $z_0(y)$ $= [x_0, y] = y(x_0)$ for every y in $\mathcal{V}'$; the correspondence $z_0 \rightleftarrows x_0$ between $\mathcal{V}''$ and $\mathcal{V}$ is an isomorphism.*

The correspondence described in this statement is called the *natural correspondence* between $\mathcal{V}''$ and $\mathcal{V}$.

PROOF. Let us view the correspondence from the standpoint of going from $\mathcal{V}$ to $\mathcal{V}''$; in other words, to every x_0 in $\mathcal{V}$ we make correspond a vector z_0 in $\mathcal{V}''$ defined by $z_0(y) = y(x_0)$ for every y in $\mathcal{V}'$. Since $[x, y]$ depends linearly on x, the transformation $x_0 \rightarrow z_0$ is linear. We shall show that this transformation is one-to-one, as far as it goes. We assert, in other words, that if x_1 and x_2 are in $\mathcal{V}$, and if z_1 and z_2 are the corresponding vectors in $\mathcal{V}''$ (so that $z_1(y) = [x_1, y]$ and $z_2(y) = [x_2, y]$ for all y in $\mathcal{V}'$), and if $z_1 = z_2$, then $x_1 = x_2$. To say that $z_1 = z_2$ means that $[x_1, y] = [x_2, y]$ for every y in $\mathcal{V}'$; the desired conclusion follows from § 15, Theorem 3.

The last two paragraphs together show that the set of those linear functionals z on $\mathcal{V}'$ (that is, elements of $\mathcal{V}''$) that do have the desired form (that is, $z(y)$ is identically equal to $[x, y]$ for a suitable x in $\mathcal{V}$) is a subspace of $\mathcal{V}''$ which is isomorphic to $\mathcal{V}$ and which is, therefore, n-dimensional. But the n-dimensionality of $\mathcal{V}$ implies that of $\mathcal{V}'$, which in turn implies that $\mathcal{V}''$ is n-dimensional. It follows that $\mathcal{V}''$ must coincide with the n-dimensional subspace just described, and the proof of the theorem is complete.

It is important to observe that the theorem shows not only that $\mathcal{V}$ and $\mathcal{V}''$ are isomorphic—this much is trivial from the fact that they have the same dimension—but that the natural correspondence is an isomorphism. This property of vector spaces is called *reflexivity*; every finite-dimensional vector space is reflexive.

It is frequently convenient to be mildly sloppy about $\mathcal{V}''$: for finite-dimensional vector spaces we shall identify $\mathcal{V}''$ with $\mathcal{V}$ (by the natural isomorphism), and we shall say that the element z_0 of $\mathcal{V}''$ is the *same* as the element x_0 of $\mathcal{V}$ whenever $z_0(y) = [x_0, y]$ for all y in $\mathcal{V}'$. In this language it is very easy to express the relation between a basis $\mathfrak{X}$, in $\mathcal{V}$, and the dual basis of its dual basis, in $\mathcal{V}''$; the symmetry of the relation $[x_i, y_j] = \delta_{ij}$ shows that $\mathfrak{X}'' = \mathfrak{X}$.

§ 17. Annihilators

DEFINITION. The *annihilator* S^0 of any subset S of a vector space $\mathcal{U}$ (S need not be a subspace) is the set of all vectors y in $\mathcal{U}'$ such that $[x, y]$ is identically zero for all x in S.

Thus $\theta^0 = \mathcal{U}'$ and $\mathcal{U}^0 = \theta$ ($\subset \mathcal{U}'$). If $\mathcal{U}$ is finite-dimensional and S contains a non-zero vector, so that $S \neq \theta$, then § 15, Theorem 3 shows that $S^0 \neq \mathcal{U}'$.

THEOREM 1. *If $\mathfrak{M}$ is an m-dimensional subspace of an n-dimensional vector space $\mathcal{U}$, then $\mathfrak{M}^0$ is an $(n - m)$-dimensional subspace of $\mathcal{U}'$.*

PROOF. We leave it to the reader to verify that $\mathfrak{M}^0$ (in fact S^0, for an arbitrary S) is always a subspace; we shall prove only the statement concerning the dimension of $\mathfrak{M}^0$.

Let $\mathfrak{X} = \{x_1, \cdots, x_n\}$ be a basis in $\mathcal{U}$ whose first m elements are in $\mathfrak{M}$ (and form therefore a basis for $\mathfrak{M}$); let $\mathfrak{X}' = \{y_1, \cdots, y_n\}$ be the dual basis in $\mathcal{U}'$. We denote by $\mathfrak{N}$ the subspace (in $\mathcal{U}'$) spanned by $y_{m+1}, \cdots, y_n$; clearly $\mathfrak{N}$ has dimension $n - m$. We shall prove that $\mathfrak{M}^0 = \mathfrak{N}$.

If x is any vector in $\mathfrak{M}$, then x is a linear combination of $x_1, \cdots, x_m$,

$$x = \sum_{i=1}^{m} \xi_i x_i,$$

and, for any $j = m + 1, \cdots, n$, we have

$$[x, y_j] = \sum_{i=1}^{m} \xi_i [x_i, y_j] = 0.$$

In other words, y_j is in $\mathfrak{M}^0$ for $j = m + 1, \cdots, n$; it follows that $\mathfrak{N}$ is contained in $\mathfrak{M}^0$,

$$\mathfrak{N} \subset \mathfrak{M}^0.$$

Suppose, on the other hand, that y is any element of $\mathfrak{M}^0$. Since y, being in $\mathcal{U}'$, is a linear combination of the basis vectors $y_1, \cdots, y_n$, we may write

$$y = \sum_{j=1}^{n} \eta_j y_j.$$

Since, by assumption, y is in $\mathfrak{M}^0$, we have, for every $i = 1, \cdots, m$,

$$0 = [x_i, y] = \sum_{j=1}^{n} \eta_j [x_i, y_j] = \eta_i;$$

in other words, y is a linear combination of $y_{m+1}, \cdots, y_n$. This proves that y is in $\mathfrak{N}$, and consequently that

$$\mathfrak{M}^0 \subset \mathfrak{N},$$

and the theorem follows.

THEOREM 2. *If $\mathfrak{M}$ is a subspace in a finite-dimensional vector space $\mathcal{V}$, then $\mathfrak{M}^{00}$ $(= (\mathfrak{M}^0)^0) = \mathfrak{M}$.*

PROOF. Observe that we use here the convention, established at the end of § 16, that identifies $\mathcal{V}$ and $\mathcal{V}''$. By definition, $\mathfrak{M}^{00}$ is the set of all vectors x such that $[x, y] = 0$ for all y in $\mathfrak{M}^0$. Since, by the definition of $\mathfrak{M}^0$, $[x, y] = 0$ for all x in $\mathfrak{M}$ and all y in $\mathfrak{M}^0$, it follows that $\mathfrak{M} \subset \mathfrak{M}^{00}$. The desired conclusion now follows from a dimension argument. Let $\mathfrak{M}$ be m-dimensional; then the dimension of $\mathfrak{M}^0$ is $n - m$, and that of $\mathfrak{M}^{00}$ is $n - (n - m) = m$. Hence $\mathfrak{M} = \mathfrak{M}^{00}$, as was to be proved.

EXERCISES

1. Define a non-zero linear functional y on $\mathcal{C}^3$ such that if $x_1 = (1, 1, 1)$ and $x_2 = (1, 1, -1)$, then $[x_1, y] = [x_2, y] = 0$.

2. The vectors $x_1 = (1, 1, 1)$, $x_2 = (1, 1, -1)$, and $x_3 = (1, -1, -1)$ form a basis of $\mathcal{C}^3$. If $\{y_1, y_2, y_3\}$ is the dual basis, and if $x = (0, 1, 0)$, find $[x, y_1]$, $[x, y_2]$, and $[x, y_3]$.

3. Prove that if y is a linear functional on an n-dimensional vector space $\mathcal{V}$, then the set of all those vectors x for which $[x, y] = 0$ is a subspace of $\mathcal{V}$; what is the dimension of that subspace?

4. If $y(x) = \xi_1 + \xi_2 + \xi_3$ whenever $x = (\xi_1, \xi_2, \xi_3)$ is a vector in $\mathcal{C}^3$, then y is a linear functional on $\mathcal{C}^3$; find a basis of the subspace consisting of all those vectors x for which $[x, y] = 0$.

5. Prove that if $m < n$, and if $y_1, \cdots, y_m$ are linear functionals on an n-dimensional vector space $\mathcal{V}$, then there exists a non-zero vector x in $\mathcal{V}$ such that $[x, y_j] = 0$ for $j = 1, \cdots, m$. What does this result say about the solutions of linear equations?

6. Suppose that $m < n$ and that $y_1, \cdots, y_m$ are linear functionals on an n-dimensional vector space $\mathcal{V}$. Under what conditions on the scalars $\alpha_1, \cdots, \alpha_m$ is it true that there exists a vector x in $\mathcal{V}$ such that $[x, y_j] = \alpha_j$ for $j = 1, \cdots, m$? What does this result say about the solutions of linear equations?

7. If $\mathcal{V}$ is an n-dimensional vector space over a finite field, and if $0 \leqq m \leqq n$ then the number of m-dimensional subspaces of $\mathcal{V}$ is the same as the number of $(n - m)$-dimensional subspaces.

8. (a) Prove that if $\mathcal{S}$ is any subset of a finite-dimensional vector space, then $\mathcal{S}^{00}$ coincides with the subspace spanned by $\mathcal{S}$.

(b) If $\mathcal{S}$ and $\mathcal{J}$ are subsets of a vector space, and if $\mathcal{S} \subset \mathcal{J}$, then $\mathcal{J}^0 \subset \mathcal{S}^0$.

(c) If $\mathfrak{M}$ and $\mathfrak{N}$ are subspaces of a finite-dimensional vector space, then $(\mathfrak{M} \cap \mathfrak{N})^0 = \mathfrak{M}^0 + \mathfrak{N}^0$ and $(\mathfrak{M} + \mathfrak{N})^0 = \mathfrak{M}^0 \cap \mathfrak{N}^0$. (Hint: make repeated use of (b) and of § 17, Theorem 2.)

(d) Is the conclusion of (c) valid for not necessarily finite-dimensional vector spaces?

9. This exercise is concerned with vector spaces that need not be finite-dimensional; most of its parts (but not all) depend on the sort of transfinite reasoning that is needed to prove that every vector space has a basis (cf. § 7, Ex. 11).

(a) Suppose that f and g are scalar-valued functions defined on a set $\mathfrak{X}$; if α and β are scalars write $h = \alpha f + \beta g$ for the function defined by $h(x) = \alpha f(x) + \beta g(x)$ for all x in $\mathfrak{X}$. The set of all such functions is a vector space with respect to this definition of the linear operations, and the same is true of the set of all finitely non-zero functions. (A function f on $\mathfrak{X}$ is *finitely non-zero* if the set of those elements x of $\mathfrak{X}$ for which $f(x) \neq 0$ is finite.)

(b) Every vector space is isomorphic to the set of all finitely non-zero functions on some set.

(c) If $\mathcal{V}$ is a vector space with basis $\mathfrak{X}$, and if f is a scalar-valued function defined on the set $\mathfrak{X}$, then there exists a unique linear functional y on $\mathcal{V}$ such that $[x, y] = f(x)$ for all x in $\mathfrak{X}$.

(d) Use (a), (b), and (c) to conclude that every vector space $\mathcal{V}$ is isomorphic to a subspace of $\mathcal{V}'$.

(e) Which vector spaces are isomorphic to their own duals?

(f) If $\mathcal{Y}$ is a linearly independent subset of a vector space $\mathcal{V}$, then there exists a basis of $\mathcal{V}$ containing $\mathcal{Y}$. (Compare this result with the theorem of § 7.)

(g) If $\mathfrak{X}$ is a set and if y is an element of $\mathfrak{X}$, write f_y for the scalar-valued function defined on $\mathfrak{X}$ by writing $f_y(x) = 1$ or 0 according as $x = y$ or $x \neq y$. Let $\mathcal{Y}$ be the set of all functions f_y together with the function g defined by $g(x) = 1$ for all x in $\mathfrak{X}$. Prove that if $\mathfrak{X}$ is infinite, then $\mathcal{Y}$ is a linearly independent subset of the vector space of all scalar-valued functions on $\mathfrak{X}$.

(h) The natural correspondence from $\mathcal{V}$ to $\mathcal{V}''$ is defined for all vector spaces (not only for the finite-dimensional ones); if x_0 is in $\mathcal{V}$, define the corresponding element z_0 of $\mathcal{V}''$ by writing $z_0(y) = [x_0, y]$ for all y in $\mathcal{V}'$. Prove that if $\mathcal{V}$ is reflexive (i.e., if every z_0 in $\mathcal{V}''$ can be obtained in this manner by a suitable choice of x_0), then $\mathcal{V}$ is finite-dimensional. (Hint: represent $\mathcal{V}'$ as the set of all scalar-valued functions on some set, and then use (g), (f), and (c) to construct an element of $\mathcal{V}''$ that is not induced by an element of $\mathcal{V}$.)

Warning: the assertion that a vector space is reflexive if and only if it is finite-dimensional would shock most of the experts in the subject. The reason is that the customary and fruitful generalization of the concept of reflexivity to infinite-dimensional spaces is not the simple-minded one given in (h).

§ 18. Direct sums

We shall study several important general methods of making new vector spaces out of old ones; in this section we begin by studying the easiest one.

DEFINITION. If $\mathcal{U}$ and $\mathcal{V}$ are vector spaces (over the same field), their *direct sum* is the vector space $\mathcal{W}$ (denoted by $\mathcal{U} \oplus \mathcal{V}$) whose elements are all the ordered pairs $\langle x, y \rangle$ with x in $\mathcal{U}$ and y in $\mathcal{V}$, with the linear operations defined by

$$\alpha_1 \langle x_1, y_1 \rangle + \alpha_2 \langle x_2, y_2 \rangle = \langle \alpha_1 x_1 + \alpha_2 x_2, \alpha_1 y_1 + \alpha_2 y_2 \rangle.$$

We observe that the formation of the direct sum is analogous to the way in which the plane is constructed from its two coordinate axes.

We proceed to investigate the relation of this notion to some of our earlier ones.

The set of all vectors (in $\mathcal{W}$) of the form $\langle x, 0 \rangle$ is a subspace of $\mathcal{W}$; the correspondence $\langle x, 0 \rangle \rightleftarrows x$ shows that this subspace is isomorphic to $\mathcal{U}$. It is convenient, once more, to indulge in a logical inaccuracy and, identifying x and $\langle x, 0 \rangle$, to speak of $\mathcal{U}$ as a subspace of $\mathcal{W}$. Similarly, of course, the vectors y of $\mathcal{V}$ may be identified with the vectors of the form $\langle 0, y \rangle$ in $\mathcal{W}$, and we may consider $\mathcal{V}$ as a subspace of $\mathcal{W}$. This terminology is, to be sure, not quite exact, but the logical difficulty is much easier to get around here than it was in the case of the second dual space. We could have defined the direct sum of $\mathcal{U}$ and $\mathcal{V}$ (at least in the case in which $\mathcal{U}$ and $\mathcal{V}$ have no non-zero vectors in common) as the set consisting of all x's in $\mathcal{U}$, all y's in $\mathcal{V}$, and all those pairs $\langle x, y \rangle$ for which $x \neq 0$ and $y \neq 0$. This definition yields a theory analogous in every detail to the one we shall develop, but it makes it a nuisance to prove theorems because of the case distinctions it necessitates. It is clear, however, that from the point of view of this definition $\mathcal{U}$ is actually a subset of $\mathcal{U} \oplus \mathcal{V}$. In this sense then, or in the isomorphism sense of the definition we did adopt, we raise the question: what is the relation between $\mathcal{U}$ and $\mathcal{V}$ when we consider these spaces as subspaces of the big space $\mathcal{W}$?

THEOREM. *If $\mathcal{U}$ and $\mathcal{V}$ are subspaces of a vector space $\mathcal{W}$, then the following three conditions are equivalent.*

(1) $\mathcal{W} = \mathcal{U} \oplus \mathcal{V}$.

(2) $\mathcal{U} \cap \mathcal{V} = \mathcal{O}$ *and* $\mathcal{U} + \mathcal{V} = \mathcal{W}$ (*i.e.*, $\mathcal{U}$ *and* $\mathcal{V}$ *are complements of each other*).

(3) *Every vector z in $\mathcal{W}$ may be written in the form $z = x + y$, with x in $\mathcal{U}$ and y in $\mathcal{V}$, in one and only one way.*

PROOF. We shall prove the implications (1) $\Rightarrow$ (2) $\Rightarrow$ (3) $\Rightarrow$ (1).

(1) $\Rightarrow$ (2). We assume that $\mathcal{W} = \mathcal{U} \oplus \mathcal{V}$. If $z = \langle x, y \rangle$ lies in both $\mathcal{U}$ and $\mathcal{V}$, then $x = y = 0$, so that $z = 0$; this proves that $\mathcal{U} \cap \mathcal{V} = \mathcal{O}$. Since the representation $z = \langle x, 0 \rangle + \langle 0, y \rangle$ is valid for every z, it follows also that $\mathcal{U} + \mathcal{V} = \mathcal{W}$.

(2) $\Rightarrow$ (3). If we assume (2), so that, in particular, $\mathcal{U} + \mathcal{V} = \mathcal{W}$, then it is clear that every z in $\mathcal{W}$ has the desired representation, $z = x + y$. To prove uniqueness, we assume that $z = x_1 + y_1$ and $z = x_2 + y_2$, with x_1 and x_2 in $\mathcal{U}$ and y_1 and y_2 in $\mathcal{V}$. Since $x_1 + y_1 = x_2 + y_2$, it follows that $x_1 - x_2 = y_2 - y_1$. Since the left member of this last equation is in $\mathcal{U}$ and the right member is in $\mathcal{V}$, the disjointness of $\mathcal{U}$ and $\mathcal{V}$ implies that $x_1 = x_2$ and $y_1 = y_2$.

(3) $\Rightarrow$ (1). This implication is practically indistinguishable from the definition of direct sum. If we form the direct sum $\mathcal{U} \oplus \mathcal{V}$, and then

identify $\langle x, 0 \rangle$ and $\langle 0, y \rangle$ with x and y respectively, we are committed to identifying the sum $\langle x, y \rangle = \langle x, 0 \rangle + \langle 0, y \rangle$ with what we are assuming to be the general element $z = x + y$ of $\mathcal{W}$; from the hypothesis that the representation of z in the form $x + y$ is unique we conclude that the correspondence between $\langle x, 0 \rangle$ and x (and also between $\langle 0, y \rangle$ and y) is one-to-one.

If two subspaces $\mathcal{U}$ and $\mathcal{V}$ in a vector space $\mathcal{W}$ are disjoint and span $\mathcal{W}$ (that is, if they satisfy (2)), it is usual to say that $\mathcal{W}$ is the *internal direct sum* of $\mathcal{U}$ and $\mathcal{V}$; symbolically, as before, $\mathcal{W} = \mathcal{U} \oplus \mathcal{V}$. If we want to emphasize the distinction between this concept and the one defined before, we describe the earlier one by saying that $\mathcal{W}$ is the *external direct sum* of $\mathcal{U}$ and $\mathcal{V}$. In view of the natural isomorphisms discussed above, and, especially, in view of the preceding theorem, the distinction is more pedantic than conceptual. In accordance with our identification convention, we shall usually ignore it.

§ 19. Dimension of a direct sum

What can be said about the dimension of a direct sum? If $\mathcal{U}$ is n-dimensional, $\mathcal{V}$ is m-dimensional, and $\mathcal{W} = \mathcal{U} \oplus \mathcal{V}$, what is the dimension of $\mathcal{W}$? This question is easy to answer.

THEOREM 1. *The dimension of a direct sum is the sum of the dimensions of its summands.*

PROOF. We assert that if $\{x_1, \cdots, x_n\}$ is a basis in $\mathcal{U}$, and if $\{y_1, \cdots, y_m\}$ is a basis in $\mathcal{V}$, then the set $\{x_1, \cdots, x_n, y_1, \cdots, y_m\}$ (or, more precisely, the set $\{\langle x_1, 0 \rangle, \cdots, \langle x_n, 0 \rangle, \langle 0, y_1 \rangle, \cdots, \langle 0, y_m \rangle\}$) is a basis in $\mathcal{W}$. The easiest proof of this assertion is to use the implication $(1) \Rightarrow (3)$ from the theorem of the preceding section. Since every z in $\mathcal{W}$ may be written in the form $z = x + y$, where x is a linear combination of $x_1, \cdots, x_n$ and y is a linear combination of $y_1, \cdots, y_m$, it follows that our set does indeed span $\mathcal{W}$. To show that the set is also linearly independent, suppose that

$$\alpha_1 x_1 + \cdots + \alpha_n x_n + \beta_1 y_1 + \cdots + \beta_m y_m = 0.$$

The uniqueness of the representation of 0 in the form $x + y$ implies that

$$\alpha_1 x_1 + \cdots + \alpha_n x_n = \beta_1 y_1 + \cdots + \beta_m y_m = 0,$$

and hence the linear independence of the x's and of the y's implies that

$$\alpha_1 = \cdots = \alpha_n = \beta_1 = \cdots = \beta_m = 0.$$

THEOREM 2. *If $\mathcal{W}$ is any $(n + m)$-dimensional vector space, and if $\mathcal{U}$ is any n-dimensional subspace of $\mathcal{W}$, then there exists an m-dimensional subspace $\mathcal{V}$ in $\mathcal{W}$ such that $\mathcal{W} = \mathcal{U} \oplus \mathcal{V}$.*

PROOF. Let $\{x_1, \cdots, x_n\}$ be any basis in $\mathcal{U}$; by the theorem of § 7 we may find a set $\{y_1, \cdots, y_m\}$ of vectors in $\mathcal{W}$ with the property that $\{x_1, \cdots, x_n, y_1, \cdots, y_m\}$ is a basis in $\mathcal{W}$. Let $\mathcal{V}$ be the subspace spanned by $y_1, \cdots, y_m$; we omit the verification that $\mathcal{W} = \mathcal{U} \oplus \mathcal{V}$.

Theorem 2 says that every subspace of a finite-dimensional vector space has a complement.

§ 20. Dual of a direct sum

In most of what follows we shall view the notion of direct sum as defined for subspaces of a vector space $\mathcal{V}$; this avoids the fuss with the identification convention of § 18, and it turns out, incidentally, to be the more useful concept for our later work. We conclude, for the present, our study of direct sums, by observing the simple relation connecting dual spaces, annihilators, and direct sums. To emphasize our present view of direct summation, we return to the letters of our earlier notation.

THEOREM. *If $\mathfrak{M}$ and $\mathfrak{N}$ are subspaces of a vector space $\mathcal{V}$, and if $\mathcal{V} = \mathfrak{M} \oplus \mathfrak{N}$, then $\mathfrak{M}'$ is isomorphic to $\mathfrak{N}^0$ and $\mathfrak{N}'$ to $\mathfrak{M}^0$, and $\mathcal{V}' = \mathfrak{M}^0 \oplus \mathfrak{N}^0$.*

PROOF. To simplify the notation we shall use, throughout this proof, x, x', and x^0 for elements of $\mathfrak{M}$, $\mathfrak{M}'$, and $\mathfrak{M}^0$, respectively, and we reserve, similarly, the letters y for $\mathfrak{N}$ and z for $\mathcal{V}$. (This notation is not meant to suggest that there is any particular relation between, say, the vectors x in $\mathfrak{M}$ and the vectors x' in $\mathfrak{M}'$.)

If z' belongs to both $\mathfrak{M}^0$ and $\mathfrak{N}^0$, i.e., if $z'(x) = z'(y) = 0$ for all x and y, then $z'(z) = z'(x + y) = 0$ for all z; this implies that $\mathfrak{M}^0$ and $\mathfrak{N}^0$ are disjoint. If, moreover, z' is any vector in $\mathcal{V}'$, and if $z = x + y$, we write $x^0(z) = z'(y)$ and $y^0(z) = z'(x)$. It is easy to see that the functions x^0 and y^0 thus defined are linear functionals on $\mathcal{V}$ (i.e., elements of $\mathcal{V}'$) belonging to $\mathfrak{M}^0$ and $\mathfrak{N}^0$ respectively; since $z' = x^0 + y^0$, it follows that $\mathcal{V}'$ is indeed the direct sum of $\mathfrak{M}^0$ and $\mathfrak{N}^0$.

To establish the asserted isomorphisms, we make correspond to every x^0 a y' in $\mathfrak{N}'$ defined by $y'(y) = x^0(y)$. We leave to the reader the routine verification that the correspondence $x^0 \rightarrow y'$ is linear and one-to-one, and therefore an isomorphism between $\mathfrak{M}^0$ and $\mathfrak{N}'$; the corresponding result for $\mathfrak{N}^0$ and $\mathfrak{M}'$ follows from symmetry by interchanging x and y. (Observe that for finite-dimensional vector spaces the mere existence of an isomorphism between, say, $\mathfrak{M}^0$ and $\mathfrak{N}'$ is trivial from a dimension argument; indeed, the dimensions of both $\mathfrak{M}^0$ and $\mathfrak{N}'$ are equal to the dimension of $\mathfrak{N}$.)

We remark, concerning our entire presentation of the theory of direct sums, that there is nothing magic about the number two; we could have defined the direct sum of any finite number of vector spaces, and we could have proved the obvious analogues of all the theorems of the last three sections, with only the notation becoming more complicated. We serve warning that we shall use this remark later and treat the theorems it implies as if we had proved them.

1. Suppose that x, y, u, and v are vectors in $\mathbb{C}^4$; let $\mathfrak{M}$ and $\mathfrak{N}$ be the subspaces of $\mathbb{C}^4$ spanned by $\{x, y\}$ and $\{u, v\}$ respectively. In which of the following cases is it true that $\mathbb{C}^4 = \mathfrak{M} \oplus \mathfrak{N}$?
  (a) $x = (1, 1, 0, 0)$, $\quad y = (1, 0, 1, 0)$
   $u = (0, 1, 0, 1)$, $\quad v = (0, 0, 1, 1)$.
  (b) $x = (-1, 1, 1, 0)$, $y = (0, 1, -1, 1)$
   $u = (1, 0, 0, 0)$, $\quad v = (0, 0, 0, 1)$.
  (c) $x = (1, 0, 0, 1)$, $\quad y = (0, 1, 1, 0)$
   $u = (1, 0, 1, 0)$, $\quad v = (0, 1, 0, 1)$.

2. If $\mathfrak{M}$ is the subspace consisting of all those vectors $(\xi_1, \cdots, \xi_n, \xi_{n+1}, \cdots, \xi_{2n})$ in $\mathbb{C}^{2n}$ for which $\xi_1 = \cdots = \xi_n = 0$, and if $\mathfrak{N}$ is the subspace of all those vectors for which $\xi_j = \xi_{n+j}$, $j = 1, \cdots, n$, then $\mathbb{C}^{2n} = \mathfrak{M} \oplus \mathfrak{N}$.

3. Construct three subspaces $\mathfrak{M}$, $\mathfrak{N}_1$, and $\mathfrak{N}_2$ of a vector space $\mathcal{V}$ so that $\mathfrak{M} \oplus \mathfrak{N}_1 = \mathfrak{M} \oplus \mathfrak{N}_2 = \mathcal{V}$ but $\mathfrak{N}_1 \neq \mathfrak{N}_2$. (Note that this means that there is no cancellation law for direct sums.) What is the geometric picture corresponding to this situation?

4. (a) If $\mathcal{U}$, $\mathcal{V}$, and $\mathcal{W}$ are vector spaces, what is the relation between $\mathcal{U} \oplus (\mathcal{V} \oplus \mathcal{W})$ and $(\mathcal{U} \oplus \mathcal{V}) \oplus \mathcal{W}$ (i.e., in what sense is the formation of direct sums an associative operation)?
  (b) In what sense is the formation of direct sums commutative?

5. (a) Three subspaces $\mathcal{L}$, $\mathfrak{M}$, and $\mathfrak{N}$ of a vector space $\mathcal{V}$ are called *independent* if each one is disjoint from the sum of the other two. Prove that a necessary and sufficient condition for $\mathcal{V} = \mathcal{L} \oplus (\mathfrak{M} \oplus \mathfrak{N})$ (and also for $\mathcal{V} = (\mathcal{L} \oplus \mathfrak{M}) \oplus \mathfrak{N}$) is that $\mathcal{L}$, $\mathfrak{M}$, and $\mathfrak{N}$ be independent and that $\mathcal{V} = \mathcal{L} + \mathfrak{M} + \mathfrak{N}$. (The subspace $\mathcal{L} + \mathfrak{M} + \mathfrak{N}$ is the set of all vectors of the form $x + y + z$, with x in $\mathcal{L}$, y in $\mathfrak{M}$, and z in $\mathfrak{N}$.)
  (b) Give an example of three subspaces of a vector space $\mathcal{V}$, such that the sum of all three is $\mathcal{V}$, such that every two of the three are disjoint, but such that the three are not independent.
  (c) Suppose that x, y, and z are elements of a vector space and that $\mathcal{L}$, $\mathfrak{M}$, and $\mathfrak{N}$ are the subspaces spanned by x, y, and z, respectively. Prove that the vectors x, y, and z are linearly independent if and only if the subspaces $\mathcal{L}$, $\mathfrak{M}$, and $\mathfrak{N}$ are independent.
  (d) Prove that three finite-dimensional subspaces are independent if and only if the sum of their dimensions is equal to the dimension of their sum.
  (e) Generalize the results (a)–(d) from three subspaces to any finite number.

§ 21. Quotient spaces

We know already that if $\mathfrak{M}$ is a subspace of a vector space $\mathcal{V}$, then there are, usually, many other subspaces $\mathfrak{N}$ in $\mathcal{V}$ such that $\mathfrak{M} \oplus \mathfrak{N} = \mathcal{V}$. There is no natural way of choosing one from among the wealth of complements of $\mathfrak{M}$. There is, however, a natural construction that associates with $\mathfrak{M}$ and $\mathcal{V}$ a new vector space that, for all practical purposes, plays the role of a complement of $\mathfrak{M}$. The theoretical advantage that the construction has over the formation of an arbitrary complement is precisely its "natural" character, i.e., the fact that it does not depend on choosing a basis, or, for that matter, on choosing anything at all.

In order to understand the construction it is a good idea to keep a picture in mind. Suppose, for instance, that $\mathcal{V} = \mathcal{R}^2$ (the real coordinate plane) and that $\mathfrak{M}$ consists of all those vectors (ξ_1, ξ_2) for which $\xi_2 = 0$ (the horizontal axis). Each complement of $\mathfrak{M}$ is a line (other than the horizontal axis) through the origin. Observe that each such complement has the property that it intersects every horizontal line in exactly one point. The idea of the construction we shall describe is to make a vector space out of the set of all horizontal lines.

We begin by using $\mathfrak{M}$ to single out certain subsets of $\mathcal{V}$. (We are back in the general case now.) If x is an arbitrary vector in $\mathcal{V}$, we write $x + \mathfrak{M}$ for the set of all sums $x + y$ with y in $\mathfrak{M}$; each set of the form $x + \mathfrak{M}$ is called a *coset* of $\mathfrak{M}$. (In the case of the plane-line example above, the cosets are the horizontal lines.) Note that one and the same coset can arise from two different vectors, i.e., that even if $x \neq y$, it is possible that $x + \mathfrak{M} = y + \mathfrak{M}$. It makes good sense, just the same, to speak of a coset, say $\mathcal{K}$, of $\mathfrak{M}$, without specifying which element (or elements) $\mathcal{K}$ comes from; to say that $\mathcal{K}$ is a coset (of $\mathfrak{M}$) means simply that there is at least one x such that $\mathcal{K} = x + \mathfrak{M}$.

If $\mathcal{K}$ and $\mathcal{K}$ are cosets (of $\mathfrak{M}$), we write $\mathcal{K} + \mathcal{K}$ for the set of all sums $u + v$ with u in $\mathcal{K}$ and v in $\mathcal{K}$; we assert that $\mathcal{K} + \mathcal{K}$ is also a coset of $\mathfrak{M}$. Indeed, if $\mathcal{K} = x + \mathfrak{M}$ and $\mathcal{K} = y + \mathfrak{M}$, then every element of $\mathcal{K} + \mathcal{K}$ belongs to the coset $(x + y) + \mathfrak{M}$ (note that $\mathfrak{M} + \mathfrak{M} = \mathfrak{M}$), and, conversely, every element of $(x + y) + \mathfrak{M}$ is in $\mathcal{K} + \mathcal{K}$. (If, for instance, z is in $\mathfrak{M}$, then $(x + y) + z = (x + z) + (y + 0)$.) In other words, $\mathcal{K} + \mathcal{K} = (x + y) + \mathfrak{M}$, so that $\mathcal{K} + \mathcal{K}$ is a coset, as asserted. We leave to the reader the verification that coset addition is commutative and associative. The coset $\mathfrak{M}$ (i.e., $0 + \mathfrak{M}$) is such that $\mathcal{K} + \mathfrak{M} = \mathcal{K}$ for every coset $\mathcal{K}$, and, moreover, $\mathfrak{M}$ is the only coset with this property. (If $(x + \mathfrak{M}) + (y + \mathfrak{M}) = x + \mathfrak{M}$, then $x + \mathfrak{M}$ contains $x + y$, so that $x + y = x + u$ for some u in $\mathfrak{M}$; this implies that y is in $\mathfrak{M}$, and hence that $y + \mathfrak{M} = \mathfrak{M}$.) If $\mathcal{K}$ is a coset, then the set consisting of all the vectors $-u$, with u in $\mathcal{K}$, is itself a coset, which we shall denote by $-\mathfrak{K}$. The coset $-\mathfrak{K}$ is such that $\mathfrak{K} + (-\mathfrak{K}) = \mathfrak{M}$, and, moreover, $-\mathfrak{K}$ is the only coset with this property. To sum up: the addition of cosets satisfies the axioms (A) of § 2.

If $\mathfrak{K}$ is a coset and if α is a scalar, we write $\alpha\mathfrak{K}$ for the set consisting of all the vectors αu with u in $\mathfrak{K}$ in case $\alpha \neq 0$; the coset $0 \cdot \mathfrak{K}$ is defined to be $\mathfrak{M}$. A simple verification shows that this concept of multiplication satisfies the axioms (B) and (C) of § 2.

The set of all cosets has thus been proved to be a vector space with respect to the linear operations defined above. This vector space is called the *quotient space* of $\mathcal{V}$ modulo $\mathfrak{M}$; it is denoted by $\mathcal{V}/\mathfrak{M}$.

§ 22. Dimension of a quotient space

Theorem 1. *If $\mathfrak{M}$ and $\mathfrak{N}$ are complementary subspaces of a vector space $\mathcal{V}$, then the correspondence that assigns to each vector y in $\mathfrak{N}$ the coset $y + \mathfrak{M}$ is an isomorphism between $\mathfrak{N}$ and $\mathcal{V}/\mathfrak{M}$.*

PROOF. If y_1 and y_2 are elements of $\mathfrak{N}$ such that $y_1 + \mathfrak{M} = y_2 + \mathfrak{M}$, then, in particular, y_1 belongs to $y_2 + \mathfrak{M}$, so that $y_1 = y_2 + x$ for some x in $\mathfrak{M}$. Since this means that $y_1 - y_2 = x$, and since $\mathfrak{M}$ and $\mathfrak{N}$ are disjoint, it follows that $x = 0$, and hence that $y_1 = y_2$. (Recall that $y_1 - y_2$ belongs to $\mathfrak{N}$ along with y_1 and y_2.) This argument proves that the correspondence we are studying is one-to-one, as far as it goes. To prove that it goes far enough, consider an arbitrary coset of $\mathfrak{M}$, say $z + \mathfrak{M}$. Since $\mathcal{V} = \mathfrak{N} + \mathfrak{M}$, we may write z in the form $y + x$, with x in $\mathfrak{M}$ and y in $\mathfrak{N}$; it follows (since $x + \mathfrak{M} = \mathfrak{M}$) that $z + \mathfrak{M} = y + \mathfrak{M}$. This proves that every coset of $\mathfrak{M}$ can be obtained by using an element of $\mathfrak{N}$ (and not just any old element of $\mathcal{V}$); consequently $y \to y + \mathfrak{M}$ is indeed a one-to-one correspondence between $\mathfrak{N}$ and $\mathcal{V}/\mathfrak{M}$. The linear property of the correspondence is immediate from the definition of the linear operations in $\mathcal{V}/\mathfrak{M}$; indeed, we have

$$(\alpha_1 y_1 + \alpha_2 y_2) + \mathfrak{M} = \alpha_1(y_1 + \mathfrak{M}) + \alpha_2(y_2 + \mathfrak{M}).$$

Theorem 2. *If $\mathfrak{M}$ is an m-dimensional subspace of an n-dimensional vector space $\mathcal{V}$, then $\mathcal{V}/\mathfrak{M}$ has dimension $n - m$.*

PROOF. Use § 19, Theorem 2 to find a subspace $\mathfrak{N}$ so that $\mathfrak{M} \oplus \mathfrak{N} = \mathcal{V}$. The space $\mathfrak{N}$ has dimension $n - m$ (by § 19, Theorem 1), and it is isomorphic to $\mathcal{V}/\mathfrak{M}$ (by Theorem 1 above).

There are more topics in the theory of quotient spaces that we could discuss (such as their relation to dual spaces and annihilators). Since, however, most such topics are hardly more than exercises, involving the use of techniques already at our disposal, we turn instead to some new and non-obvious ways of manufacturing useful vector spaces.

EXERCISES

1. Consider the quotient spaces obtained by reducing the space $\mathcal{O}$ of polynomials modulo various subspaces. If $\mathfrak{M} = \mathcal{O}_n$, is $\mathcal{O}/\mathfrak{M}$ finite-dimensional? What if $\mathfrak{M}$ is the subspace consisting of all even polynomials? What if $\mathfrak{M}$ is the subspace consisting of all polynomials divisible by x_n (where $x_n(t) = t^n$)?

2. If S and $\mathfrak{I}$ are arbitrary subsets of a vector space (not necessarily cosets of a subspace), there is nothing to stop us from defining $S + \mathfrak{I}$ just as addition was defined for cosets, and, similarly, we may define αS (where α is a scalar). If the class of all subsets of a vector space is endowed with these "linear operations," which of the axioms of a vector space are satisfied?

3. (a) Suppose that $\mathfrak{M}$ is a subspace of a vector space $\mathcal{V}$. Two vectors x and y of $\mathcal{V}$ are *congruent* modulo $\mathfrak{M}$, in symbols $x \equiv y \ (\mathfrak{M})$, if $x - y$ is in $\mathfrak{M}$. Prove that congruence modulo $\mathfrak{M}$ is an *equivalence relation*, i.e., that it is reflexive ($x \equiv x$), symmetric (if $x \equiv y$, then $y \equiv x$), and transitive (if $x \equiv y$ and $y \equiv z$, then $x \equiv z$).
   (b) If α_1 and α_2 are scalars, and if x_1, x_2, y_1, and y_2 are vectors such that $x_1 \equiv y_1$ $(\mathfrak{M})$ and $x_2 \equiv y_2 \ (\mathfrak{M})$, then $\alpha_1 x_1 + \alpha_2 x_2 \equiv \alpha_1 y_1 + \alpha_2 y_2 \ (\mathfrak{M})$.
   (c) Congruence modulo $\mathfrak{M}$ splits $\mathcal{V}$ into equivalence classes, i.e., into sets such that two vectors belong to the same set if and only if they are congruent. Prove that a subset of $\mathcal{V}$ is an equivalence class modulo $\mathfrak{M}$ if and only if it is a coset of $\mathfrak{M}$.

4. (a) Suppose that $\mathfrak{M}$ is a subspace of a vector space $\mathcal{V}$. Corresponding to every linear functional y on $\mathcal{V}/\mathfrak{M}$ (i.e., to every element y of $(\mathcal{V}/\mathfrak{M})'$), there is a linear functional z on $\mathcal{V}$ (i.e., an element of $\mathcal{V}'$); the linear functional z is defined by $z(x) = y(x + \mathfrak{M})$. Prove that the correspondence $y \rightarrow z$ is an isomorphism between $(\mathcal{V}/\mathfrak{M})'$ and $\mathfrak{M}^0$.
   (b) Suppose that $\mathfrak{M}$ is a subspace of a vector space $\mathcal{V}$. Corresponding to every coset $y + \mathfrak{M}^0$ of $\mathfrak{M}^0$ in $\mathcal{V}'$ (i.e., to every element $\mathfrak{K}$ of $\mathcal{V}'/\mathfrak{M}^0$), there is a linear functional z on $\mathfrak{M}$ (i.e., an element z of $\mathfrak{M}'$); the linear functional z is defined by $z(x) = y(x)$. Prove that z is unambiguously determined by the coset $\mathfrak{K}$ (that is, it does not depend on the particular choice of y), and that the correspondence $\mathfrak{K} \rightarrow z$ is an isomorphism between $\mathcal{V}'/\mathfrak{M}^0$ and $\mathfrak{M}'$.

5. Given a finite-dimensional vector space $\mathcal{V}$, form the direct sum $\mathcal{W} = \mathcal{V} \oplus \mathcal{V}'$, and prove that the correspondence $\langle x, y \rangle \rightarrow \langle y, x \rangle$ is an isomorphism between $\mathcal{W}$ and $\mathcal{W}'$.

§ 23. Bilinear forms

If $\mathcal{U}$ and $\mathcal{V}$ are vector spaces (over the same field), then their direct sum $\mathcal{W} = \mathcal{U} \oplus \mathcal{V}$ is another vector space; we propose to study certain functions on $\mathcal{W}$. (For present purposes the original definition of $\mathcal{U} \oplus \mathcal{V}$, via ordered pairs, is the convenient one.) The value of such a function, say w, at an element $\langle x, y \rangle$ of $\mathcal{W}$ will be denoted by $w(x, y)$. The study of linear functions on $\mathcal{W}$ is no longer of much interest to us; the principal facts concerning them were discussed in § 20. The functions we want to consider now are the bilinear ones; they are, by definition, the scalar-valued functions on $\mathcal{W}$ with the property that for each fixed value of either argument they depend linearly on the other argument. More precisely, a scalar-valued function w on $\mathcal{W}$ is a *bilinear form* (or *bilinear functional*) if

$$w(\alpha_1 x_1 + \alpha_2 x_2, y) = \alpha_1 w(x_1, y) + \alpha_2 w(x_2, y)$$

and

$$w(x, \alpha_1 y_1 + \alpha_2 y_2) = \alpha_1 w(x, y_1) + \alpha_2 w(x, y_2),$$

identically in the vectors and scalars involved.

In one special situation we have already encountered bilinear functionals. If, namely, $\mathcal{V}$ is the dual space of $\mathcal{U}$, $\mathcal{V} = \mathcal{U}'$, and if we write $w(x, y) = [x, y]$ (see § 14), then w is a bilinear functional on $\mathcal{U} \oplus \mathcal{U}'$. For an example in a more general situation, let $\mathcal{U}$ and $\mathcal{V}$ be arbitrary vector spaces (over the same field, as always), let u and v be elements of $\mathcal{U}'$ and $\mathcal{V}'$ respectively, and write $w(x, y) = u(x)v(y)$ for all x in $\mathcal{U}$ and y in $\mathcal{V}$. An even more general example is obtained by selecting a finite number of elements in $\mathcal{U}'$, say $u_1, \cdots, u_k$, selecting the same finite number of elements in $\mathcal{V}'$, say $v_1, \cdots, v_k$, and writing $w(x, y) = u_1(x)v_1(y) + \cdots + u_k(x)v_k(y)$. Which of the words, "functional" or "form," is used depends somewhat on the context and, somewhat more, on the user's whim. In this book we shall generally use "functional" with "linear" and "form" with "bilinear" (and its higher-dimensional generalizations).

If w_1 and w_2 are bilinear forms on $\mathcal{W}$, and if α_1 and α_2 are scalars, we write w for the function on $\mathcal{W}$ defined by

$$w(x, y) = \alpha_1 w_1(x, y) + \alpha_2 w_2(x, y).$$

It is easy to see that w is a bilinear form; we denote it by $\alpha_1 w_1 + \alpha_2 w_2$. With this definition of the linear operations, the set of all bilinear forms on $\mathcal{W}$ is a vector space. The chief purpose of the remainder of this section is to determine (in the finite-dimensional case) how the dimension of this space depends on the dimensions of $\mathcal{U}$ and $\mathcal{V}$.

THEOREM 1. *If $\mathcal{U}$ is an n-dimensional vector space with basis $\{x_1, \cdots, x_n\}$, if $\mathcal{V}$ is an m-dimensional vector space with basis $\{y_1, \cdots, y_m\}$, and if $\{\alpha_{ij}\}$ is any set of nm scalars $(i = 1, \cdots, n; j = 1, \cdots, m)$, then there is one and only one bilinear form w on $\mathcal{U} \oplus \mathcal{V}$ such that $w(x_i, y_j) = \alpha_{ij}$ for all i and j.*

PROOF. If $x = \sum_i \xi_i x_i$, $y = \sum_j \eta_j y_j$, and w is a bilinear form on $\mathcal{U} \oplus \mathcal{V}$ such that $w(x_i, y_j) = \alpha_{ij}$, then

$$w(x, y) = \sum_i \sum_j \xi_i \eta_j w(x_i, y_j) = \sum_i \sum_j \xi_i \eta_j \alpha_{ij}.$$

From this equation the uniqueness of w is clear; the existence of a suitable w is proved by reading the same equation from right to left, that is, defining w by it. (Compare this result with § 15, Theorem 1.)

THEOREM 2. *If $\mathfrak{U}$ is an n-dimensional vector space with basis $\{x_1, \cdots, x_n\}$, and if $\mathfrak{V}$ is an m-dimensional vector space with basis $\{y_1, \cdots, y_m\}$, then there is a uniquely determined basis $\{w_{pq}\}$ $(p = 1, \cdots, n; q = 1, \cdots, m)$ in the vector space of all bilinear forms on $\mathfrak{U} \oplus \mathfrak{V}$ with the property that $w_{pq}(x_i, x_j) = \delta_{ip}\delta_{jq}$. Consequently the dimension of the space of bilinear forms on $\mathfrak{U} \oplus \mathfrak{V}$ is the product of the dimensions of $\mathfrak{U}$ and $\mathfrak{V}$.*

PROOF. Using Theorem 1, we determine w_{pq} (for each fixed p and q) by the given condition $w_{pq}(x_i, y_j) = \delta_{ip}\delta_{jq}$. The bilinear forms so determined are linearly independent, since

$$\sum_p \sum_q \alpha_{pq} w_{pq} = 0$$

implies that

$$0 = \sum_p \sum_q \alpha_{pq} \delta_{ip}\delta_{jq} = \alpha_{ij}.$$

If, moreover, w is an arbitrary element of $\mathfrak{W}$, and if $w(x_i, y_j) = \alpha_{ij}$, then $w = \sum_p \sum_q \alpha_{pq} w_{pq}$. Indeed, if $x = \sum_i \xi_i x_i$ and $y = \sum_j \eta_j y_j$, then

$$w_{pq}(x, y) = \sum_i \sum_j \xi_i \eta_j \delta_{ip}\delta_{jq} = \xi_p \eta_q,$$

and, consequently,

$$w(x, y) = \sum_i \sum_j \xi_i \eta_j \alpha_{ij} = \sum_p \sum_q \alpha_{pq} w_{pq}(x, y).$$

It follows that the w_{pq} form a basis in the space of bilinear forms; this completes the proof of the theorem. (Compare this result with § 15, Theorem 2.)

1. (a) If w is a bilinear form on $\mathfrak{R}^n \oplus \mathfrak{R}^n$, then there exist scalars α_{ij}, $i, j = 1, \cdots$, n, such that if $x = (\xi_1, \cdots, \xi_n)$ and $y = (\eta_1, \cdots, \eta_n)$, then $w(x, y) = \sum_i \sum_j \alpha_{ij} \xi_i \eta_j$. The scalars α_{ij} are uniquely determined by w.

(b) If z is a linear functional on the space of all bilinear forms on $\mathfrak{R}^n \oplus \mathfrak{R}^n$, then there exist scalars β_{ij} such that (in the notation of (a)) $z(w) = \sum_i \sum_j \alpha_{ij}\beta_{ij}$ for every w. The scalars β_{ij} are uniquely determined by z.

2. A bilinear form w on $\mathfrak{U} \oplus \mathfrak{V}$ is *degenerate* if, as a function of one of its two arguments, it vanishes identically for some non-zero value of its other argument; otherwise it is *non-degenerate*.

(a) Give an example of a degenerate bilinear form (not identically zero) on $\mathfrak{C}^2 \oplus \mathfrak{C}^2$.

(b) Give an example of a non-degenerate bilinear form on $\mathfrak{C}^2 \oplus \mathfrak{C}^2$.

3. If w is a bilinear form on $\mathfrak{U} \oplus \mathfrak{V}$, if y_0 is in $\mathfrak{V}$, and if a function y is defined on $\mathfrak{U}$ by $y(x) = w(x, y_0)$, then y is a linear functional on $\mathfrak{U}$. Is it true that if w is non-degenerate, then every linear functional on $\mathfrak{U}$ can be obtained this way (by a suitable choice of y_0)?

4. Suppose that for each x and y in $\mathcal{O}_n$ the function w is defined by

(a) $w(x, y) = \displaystyle\int_0^1 x(t)y(t)\, dt,$

(b) $w(x, y) = x(1) + y(1),$
(c) $w(x, y) = x(1) \cdot y(1),$

(d) $w(x, y) = x(1) \left(\dfrac{dy}{dt}\right)_{t=1}.$

In which of these cases is w a bilinear form on $\mathcal{O}_n \oplus \mathcal{O}_n$? In which cases is it non-degenerate?

5. Does there exist a vector space $\mathfrak{V}$ and a bilinear form w on $\mathfrak{V} \oplus \mathfrak{V}$ such that w is not identically zero but $w(x, x) = 0$ for every x in $\mathfrak{V}$?

6. (a) A bilinear form w on $\mathfrak{V} \oplus \mathfrak{V}$ is *symmetric* if $w(x, y) = w(y, x)$ for all x and y. A *quadratic form* on $\mathfrak{V}$ is a function q on $\mathfrak{V}$ obtained from a bilinear form w by writing $q(x) = w(x, x)$. Prove that if the characteristic of the underlying scalar field is different from 2, then every symmetric bilinear form is uniquely determined by the corresponding quadratic form. What happens if the characteristic is 2?
(b) Can a non-symmetric bilinear form define the same quadratic form as a symmetric one?

§ 24. Tensor products

In this section we shall describe a new method of putting two vector spaces together to make a third, namely, the formation of their tensor product. Although we shall have relatively little occasion to make use of tensor products in this book, their theory is closely allied to some of the subjects we shall treat, and it is useful in other related parts of mathematics, such as the theory of group representations and the tensor calculus. The notion is essentially more complicated than that of direct sum; we shall therefore begin by giving some examples of what a tensor product should be, and the study of these examples will guide us in laying down the definition.

Let $\mathfrak{U}$ be the set of all polynomials in one variable s, with, say, complex coefficients; let $\mathfrak{V}$ be the set of all polynomials in another variable t; and, finally, let $\mathfrak{W}$ be the set of all polynomials in the two variables s and t. With respect to the obvious definitions of the linear operations, $\mathfrak{U}$, $\mathfrak{V}$, and $\mathfrak{W}$ are all complex vector spaces; in this case we should like to call $\mathfrak{W}$, or something like it, the tensor product of $\mathfrak{U}$ and $\mathfrak{V}$. One reason for this terminology is that if we take any x in $\mathfrak{U}$ and any y in $\mathfrak{V}$, we may form their product, that is, the element z of $\mathfrak{W}$ defined by $z(s, t) = x(s)y(t)$.

(This is the ordinary product of two polynomials. Here, as before, we are doggedly ignoring the irrelevant fact that we may even multiply together two elements of $\mathfrak{U}$, that is, that the product of two polynomials in the same variable is another polynomial in that variable. Vector spaces in which a decent concept of multiplication is defined are called *algebras*, and their study, as such, lies outside the scope of this book.)

In the preceding example we considered vector spaces whose elements are functions. We may, if we wish, consider the simple vector space $\mathfrak{C}^n$ as a collection of functions also; the domain of definition of the functions is, in this case, a set consisting of exactly n points, say the first n (strictly) positive integers. In other words, a vector $(\xi_1, \cdots, \xi_n)$ may be considered as a function ξ whose value $\xi(i)$ is defined for $i = 1, \cdots, n$; the definition of the vector operations in $\mathfrak{C}^n$ is such that they correspond, in the new notation, to the ordinary operations performed on the functions ξ. If, simultaneously, we consider $\mathfrak{C}^m$ as the collection of functions η whose value $\eta(j)$ is defined for $j = 1, \cdots, m$, then we should like the tensor product of $\mathfrak{C}^n$ and $\mathfrak{C}^m$ to be the set of all functions ζ whose value $\zeta(i, j)$ is defined for $i = 1, \cdots, n$ and $j = 1, \cdots, m$. The tensor product, in other words, is the collection of all functions defined on a set consisting of exactly nm objects, and therefore naturally isomorphic to $\mathfrak{C}^{nm}$. This example brings out a property of tensor products—namely, the multiplicativity of dimension —that we should like to retain in the general case.

Let us now try to abstract the most important properties of these examples. The definition of direct sum was one possible rigorization of the crude intuitive idea of writing down, formally, the sum of two vectors belonging to different vector spaces. Similarly, our examples suggest that the tensor product $\mathfrak{U} \otimes \mathfrak{V}$ of two vector spaces $\mathfrak{U}$ and $\mathfrak{V}$ should be such that to every x in $\mathfrak{U}$ and y in $\mathfrak{V}$ there corresponds a "product" $z = x \otimes y$ in $\mathfrak{U} \otimes \mathfrak{V}$, in such a way that the correspondence between x and z, for each fixed y, as well as the correspondence between y and z, for each fixed x, is linear. (This means, of course, that $(\alpha_1 x_1 + \alpha_2 x_2) \otimes y$ should be equal to $\alpha_1(x_1 \otimes y) + \alpha_2(x_2 \otimes y)$, and that a similar equation should hold for $x \otimes (\alpha_1 y_1 + \alpha_2 y_2)$.) To put it more simply, $x \otimes y$ should define a bilinear (vector-valued) function of x and y.

The notion of formal multiplication suggests also that if u and v are linear functionals on $\mathfrak{U}$ and $\mathfrak{V}$ respectively, then it is their product w, defined by $w(x, y) = u(x)v(y)$, that should be in some sense the general element of the dual space $(\mathfrak{U} \otimes \mathfrak{V})'$. Observe that this product is a bilinear (scalar-valued) function of x and y.

§ 25. Product bases

After one more word of preliminary explanation we shall be ready to discuss the formal definition of tensor products. It turns out to be technically preferable to get at $\mathfrak{U} \otimes \mathfrak{V}$ indirectly, by defining it as the dual of another space; we shall make tacit use of reflexivity to obtain $\mathfrak{U} \otimes \mathfrak{V}$ itself. Since we have proved reflexivity for finite-dimensional spaces only, we shall restrict the definition to such spaces.

DEFINITION. The *tensor product* $\mathfrak{U} \otimes \mathfrak{V}$ of two finite-dimensional vector spaces $\mathfrak{U}$ and $\mathfrak{V}$ (over the same field) is the dual of the vector space of all bilinear forms on $\mathfrak{U} \oplus \mathfrak{V}$. For each pair of vectors x and y, with x in $\mathfrak{U}$ and y in $\mathfrak{V}$, the tensor product $z = x \otimes y$ of x and y is the element of $\mathfrak{U} \otimes \mathfrak{V}$ defined by $z(w) = w(x, y)$ for every bilinear form w.

This definition is one of the quickest rigorous approaches to the theory, but it does lead to some unpleasant technical complications later. Whatever its disadvantages, however, we observe that it obviously has the two desired properties: it is clear, namely, that dimension is multiplicative (see § 23, Theorem 2, and § 15, Theorem 2), and it is clear that $x \otimes y$ depends linearly on each of its factors.

Another possible (and deservedly popular) definition of tensor product is by formal products. According to that definition $\mathfrak{U} \otimes \mathfrak{V}$ is obtained by considering all symbols of the form $\sum_i \alpha_i(x_i \otimes y_i)$, and, within the set of such symbols, making the identifications demanded by the linearity of the vector operations and the bilinearity of tensor multiplication. (For the purist: in this definition $x \otimes y$ stands merely for the ordered pair of x and y; the multiplication sign is just a reminder of what to expect.) Neither definition is simple; we adopted the one we gave because it seemed more in keeping with the spirit of the rest of the book. The main disadvantage of our definition is that it does not readily extend to the most useful generalizations of finite-dimensional vector spaces, that is, to modules and to infinite-dimensional spaces.

For the present we prove only one theorem about tensor products. The theorem is a further justification of the product terminology, and, incidentally, it is a sharpening of the assertion that dimension is multiplicative.

THEOREM. *If* $\mathfrak{X} = \{x_1, \cdots, x_n\}$ *and* $\mathfrak{Y} = \{y_1, \cdots, y_m\}$ *are bases in* $\mathfrak{U}$ *and* $\mathfrak{V}$ *respectively, then the set* $\mathfrak{Z}$ *of vectors* $z_{ij} = x_i \otimes y_j$ $(i = 1, \cdots, n;$ $j = 1, \cdots, m)$ *is a basis in* $\mathfrak{U} \otimes \mathfrak{V}$.

PROOF. Let w_{pq} be the bilinear form on $\mathfrak{U} \oplus \mathfrak{V}$ such that $w_{pq}(x_i, y_j) = \delta_{ip}\delta_{jq}$ $(i, p = 1, \cdots, n; j, q = 1, \cdots, m)$; the existence of such bilinear forms, and the fact that they constitute a basis for all bilinear forms, follow from § 23, Theorem 2. Let $\{w'_{pq}\}$ be the dual basis in $\mathfrak{U} \otimes \mathfrak{V}$, so that $[w_{ij}, w'_{pq}] = \delta_{ip}\delta_{jq}$. If $w = \sum_p \sum_q \alpha_{pq}w_{pq}$ is an arbitrary bilinear form on $\mathfrak{U} \oplus \mathfrak{V}$, then

$$w'_{ij}(w) = [w, w'_{ij}] = \sum_p \sum_q \alpha_{pq}[w_{pq}, w'_{ij}]$$

$$= \alpha_{ij} = w(x_i, y_j) = z_{ij}(w).$$

The conclusion follows from the fact that the vectors w'_{ij} do constitute a basis of $\mathfrak{U} \otimes \mathfrak{V}$.

1. If $x = (1, 1)$ and $y = (1, 1, 1)$ are vectors in $\mathfrak{R}^2$ and $\mathfrak{R}^3$ respectively, find the coordinates of $x \otimes y$ in $\mathfrak{R}^2 \otimes \mathfrak{R}^3$ with respect to the product basis $\{x_i \otimes y_j\}$, where $x_i = (\delta_{i1}, \delta_{i2})$ and $y_j = (\delta_{1j}, \delta_{2j}, \delta_{3j})$.

2. Let $\mathcal{P}_{n,m}$ be the space of all polynomials z with complex coefficients, in two variables s and t, such that either $z = 0$ or else the degree of $z(s, t)$ is $\leq m - 1$ for each fixed s and $\leq n - 1$ for each fixed t. Prove that there exists an isomorphism between $\mathcal{P}_n \otimes \mathcal{P}_m$ and $\mathcal{P}_{n,m}$ such that the element z of $\mathcal{P}_{n,m}$ that corresponds to $x \otimes y$ (x in $\mathcal{P}_n$, y in $\mathcal{P}_m$) is given by $z(s, t) = x(s)y(t)$.

3. To what extent is the formation of tensor products commutative and associative? What about the distributive law $\mathfrak{U} \otimes (\mathfrak{V} \oplus \mathfrak{W}) = (\mathfrak{U} \otimes \mathfrak{V}) \oplus (\mathfrak{U} \otimes \mathfrak{W})$?

4. If $\mathfrak{V}$ is a finite-dimensional vector space, and if x and y are in $\mathfrak{V}$, is it true that $x \otimes y = y \otimes x$?

5. (a) Suppose that $\mathfrak{V}$ is a finite-dimensional real vector space, and let $\mathfrak{U}$ be the set $\mathcal{C}$ of all complex numbers regarded as a (two-dimensional) real vector space. Form the tensor product $\mathfrak{V}^+ = \mathfrak{U} \otimes \mathfrak{V}$. Prove that there is a way of defining products of complex numbers with elements of $\mathfrak{V}^+$ so that $\alpha(x \otimes y) = \alpha x \otimes y$ whenever α and x are in $\mathcal{C}$ and y is in $\mathfrak{V}$.

(b) Prove that with respect to vector addition, and with respect to complex scalar multiplication as defined in (a), the space $\mathfrak{V}^+$ is a complex vector space.

(c) Find the dimension of the complex vector space $\mathfrak{V}^+$ in terms of the dimension of the real vector space $\mathfrak{V}$.

(d) Prove that the vector space $\mathfrak{V}$ is isomorphic to a subspace in $\mathfrak{V}^+$ (when the latter is regarded as a real vector space).

The moral of this exercise is that not only can every complex vector space be regarded as a real vector space, but, in a certain sense, the converse is true. The vector space $\mathfrak{V}^+$ is called the *complexification* of $\mathfrak{V}$.

6. If $\mathfrak{U}$ and $\mathfrak{V}$ are finite-dimensional vector spaces, what is the dual space of $\mathfrak{U}' \otimes \mathfrak{V}'$?

§ 26. Permutations

The main subject of this book is usually known as linear algebra. In the last three sections, however, the emphasis was on something called multilinear algebra. It is hard to say exactly where the dividing line is between the two subjects. Since, in any case, both are quite extensive, it would not be practical to try to stuff a detailed treatment of both into the same volume. Nor is it desirable to discuss linear algebra in its absolutely pure state; the addition of even a small part of the multilinear theory (such as is involved in the modern view of tensor products and determinants) extends the domain of applicability of the linear theory pleasantly out of proportion with the effort involved. We propose, accordingly, to continue the study of multilinear algebra; our intention is to draw a more or less straight line between what we already know and the basic facts about determinants. With that in mind, we shall devote three sections to the discussion of some simple facts about combinatorics; the connection between those facts and multilinear algebra will appear immediately after that discussion.

By a *permutation* of the integers between 1 and k (inclusive) we shall mean a one-to-one transformation that assigns to each such integer another one (or possibly the same one). To say that the transformation π is one-to-one means, of course, that if $\pi(1)$, $\cdots$, $\pi(k)$ are the integers that π assigns to 1, $\cdots$, k, respectively, then $\pi(i) = \pi(j)$ can happen only in case $i = j$. Since this implies that both the sets $\{1, \cdots, k\}$ and $\{\pi(1), \cdots, \pi(k)\}$ consist of exactly k elements, it follows that they consist of exactly the same elements. From this, in turn, we infer that a permutation π of the set $\{1, \cdots, k\}$ maps that set *onto* itself, that is, that if $1 \leq j \leq k$, then there exists at least one i (and, in fact, exactly one) such that $\pi(i) = j$. The total number of the integers under consideration, namely, k, will be held fixed throughout the following discussion.

The theory of permutations, like everything else, is best understood by staring hard at some non-trivial examples. Before presenting any examples, however, we shall first mention some of the general things that can be done with permutations; by this means the examples will illustrate not only the basic concept but also its basic properties.

If σ and τ are arbitrary permutations, a permutation (to be denoted by $\sigma\tau$) can be defined by writing

$$(\sigma\tau)(i) = \sigma(\tau i)$$

for each i. To prove that $\sigma\tau$ is indeed a permutation, observe that if $(\sigma\tau)(i) = (\sigma\tau)(j)$, then $\tau(i) = \tau(j)$ (since σ is one-to-one), and therefore $i = j$ (since τ is one-to-one). The permutation $\sigma\tau$ is called the *product* of the permutations σ and τ. Warning: the order is important. In general $\sigma\tau \neq \tau\sigma$, or, in other words, permutation multiplication is not commutative.

Multiplication of permutations is associative; that is, if π, σ, and τ are permutations, then

(1) $$(\pi\sigma)\tau = \pi(\sigma\tau).$$

To prove this, we must show that

$$((\pi\sigma)\tau)(i) = (\pi(\sigma\tau))(i)$$

for every i. The proof consists of several applications of the definition of product, as follows:

$$((\pi\sigma)\tau)(i) = (\pi\sigma)(\tau i) = \pi(\sigma(\tau(i))),$$

and

$$(\pi(\sigma\tau))(i) = \pi((\sigma\tau)(i)) = \pi(\sigma(\tau(i))).$$

In view of this result we may and shall omit parentheses in writing the product of three or more permutations. The result also enables us to prove the obvious laws of exponents. The powers of a permutation π are defined inductively by writing $\pi^1 = \pi$ and $\pi^{p+1} = \pi \cdot \pi^p$ for all $p = 1, 2, 3, \cdots$; the associative law implies that $\pi^p\pi^q = \pi^{p+q}$ and $(\pi^p)^q = \pi^{pq}$ for all p and q. Observe that any two powers of a permutation commute with each other, that is, that $\pi^p\pi^q = \pi^q\pi^p$.

The simplest permutation is the *identity* (to be denoted by ϵ); it is defined by $\epsilon(i) = i$ for each i. If π is an arbitrary permutation, then

$$(2) \qquad \epsilon\pi = \pi\epsilon = \pi,$$

or, in other words, multiplication by ϵ leaves every permutation unaffected. The proof is straightforward; for every i we have

$$(\epsilon\pi)(i) = \epsilon(\pi(i)) = \pi(i)$$

and

$$(\pi\epsilon)(i) = \pi(\epsilon(i)) = \pi(i).$$

The permutation ϵ behaves, from the point of view of multiplication, like the number 1. In analogy with the usual numerical convention, the zero-th power of every permutation π is defined by writing $\pi^0 = \epsilon$.

If π is an arbitrary permutation, then there exists a permutation (to be denoted by π^{-1}) such that

$$(3) \qquad \pi^{-1}\pi = \pi\pi^{-1} = \epsilon.$$

To define $\pi^{-1}(j)$, where, of course, $1 \leq j \leq k$, find the unique i such that $\pi(i) = j$, and write $\pi^{-1}(j) = i$; the validity of (3) is an immediate consequence of the definitions. The permutation π^{-1} is called the *inverse* of π.

Let S_k be the set of all permutations of the integers between 1 and k. What we have proved so far is that an operation of multiplication can be defined for the elements of S_k so that (1) multiplication is associative, (2) there exists an identity element, that is, an element such that multiplication by it leaves every element of S_k fixed, and (3) every element has an inverse, that is, an element whose product with the given one is the identity. A set satisfying (1)–(3) is called a *group* with respect to the concept of product that those conditions refer to; the set S_k, in particular, is called the *symmetric group* of *degree k*. Observe that the integers $1, \cdots, k$ could be replaced by any k distinct objects without affecting any of the concepts defined above; the change would be merely a notational matter.

§ 27. Cycles

A simple example of a permutation is obtained as follows: choose any two distinct integers between 1 and k, say, p and q, and write

$$\tau(p) = q,$$

$$\tau(q) = p,$$

$$\tau(i) = i \text{ whenever } i \neq p \text{ and } i \neq q.$$

The permutation τ so defined is denoted by (p, q); every permutation of this form is called a *transposition*. If τ is a transposition, then $\tau^2 = \epsilon$.

Another useful way of constructing examples is to choose p distinct integers between 1 and k, say, $i_1, \cdots, i_p$, and to write

$$\sigma(i_j) = i_{j+1} \text{ whenever } 1 \leqq j < p,$$

$$\sigma(i_p) = i_1,$$

$$\sigma(i) = i \text{ whenever } i \neq i_1, \cdots, i \neq i_p.$$

The permutation σ so defined is denoted by $(i_1, \cdots, i_p)$. If $p = 1$, then $\sigma = \epsilon$; if $p = 2$, then σ is a transposition. For any p with $1 < p \leqq k$, every permutation of the form $(i_1, \cdots, i_p)$ is called a *p-cycle*, or simply a *cycle*; the 2-cycles are exactly the transpositions. Warning: it is not assumed that $i_1 < \cdots < i_p$. If, for instance, $k = 5$ and $p = 3$, then there are twenty distinct cycles. Observe also that the notation for cycles is not unique; the symbols (1, 2, 3), (2, 3, 1), and (3, 1, 2) all denote the same permutation. Two cycles $(i_1, \cdots, i_p)$ and $(j_1, \cdots, j_q)$ are *disjoint* if none of the i's is equal to any of the j's. If σ and τ are disjoint cycles, then $\sigma\tau = \tau\sigma$, or, in other words σ and τ commute.

THEOREM 1. *Every permutation is the product of pairwise disjoint cycles.*

PROOF. If π is a permutation and if i is such that $\pi(i) \neq i$ (assume, for the moment, that $\pi \neq \epsilon$), form the sequence $(i, \pi(i), \pi^2(i), \cdots)$. Since there are only a finite number of distinct integers between 1 and k, there must exist exponents p and q $(0 \leqq p < q)$ such that $\pi^p(i) = \pi^q(i)$. The one-to-one character of π implies that $\pi^{q-p}(i) = i$, or, with an obvious change of notation, what we have proved is that there must exist a strictly positive exponent p such that $\pi^p(i) = i$. If p is selected to be the smallest exponent with this property, then the integers $i, \cdots, \pi^{p-1}(i)$ are distinct from each other. (Indeed, if $0 \leq q < r < p$ and $\pi^q(i) = \pi^r(i)$, then $\pi^{r-q}(i) = i$, contradicting the minimality of p.) It follows that $(i, \cdots, \pi^{p-1}(i))$ is a p-cycle. If there is a j between 1 and k different from each of $i, \cdots, \pi^{p-1}(i)$ and different from $\pi(j)$, we repeat the procedure that led us to this cycle, with j in place of i. We continue forming cycles in this manner as long as after each step we can still find a new integer that π does not send on itself; the product of the disjoint cycles so constructed is π. The case $\pi = \epsilon$ is covered by the rather natural agreement that a product with no factors, an "empty product," is to be interpreted as the identity permutation.

THEOREM 2. *Every cycle is a product of transpositions.*

PROOF. Suppose that σ is a p-cycle; for the sake of notational simplicity, we shall give the proof, which is perfectly general, in the special case $p = 5$. The proof itself consists of one line:

$$(i_1, i_2, i_3, i_4, i_5) = (i_1, i_5)(i_1, i_4)(i_1, i_3)(i_1, i_2).$$

A few added words of explanation might be helpful. In view of the definition of the product of permutations, the right side of the last equation operates on each integer between 1 and k from the inside out, or, perhaps more suggestively, from right to left. Thus, for example, the result of applying $(i_1, i_5)(i_1, i_4)(i_1, i_3)(i_1, i_2)$ to i_3 is calculated as follows: $(i_1, i_2)(i_3) = i_3$, $(i_1, i_3)(i_3) = i_1$, $(i_1, i_4)(i_1) = i_4$, $(i_1, i_5)(i_4) = i_4$, so that $(i_1, i_5)(i_1, i_4)(i_1, i_3)(i_1, i_2)(i_3) = i_4$.

For the sake of reference we put on record the following immediate corollary of the two preceding theorems.

THEOREM 3. *Every permutation is a product of transpositions.*

Observe that the transpositions in Theorems 2 and 3 were not asserted to be disjoint; in general they are not.

EXERCISES

1. (a) How many permutations are there in S_k?
(b) How many distinct p-cycles are there in S_k $(1 \leq p \leq k)$?

2. If σ and τ are permutations (in S_k), then $(\sigma\tau)^{-1} = \tau^{-1}\sigma^{-1}$.

3. (a) If σ and τ are permutations (in S_k), then there exists a unique permutation π such that $\sigma\pi = \tau$.
(b) If π, σ, and τ are permutations such that $\pi\sigma = \pi\tau$, then $\sigma = \tau$.

4. Give an example of a permutation that is not the product of disjoint transpositions.

5. Prove that every permutation in S_k is the product of transpositions of the form $(j, j + 1)$, where $1 \leq j < k$. Is this factorization unique?

6. Is the inverse of a cycle also a cycle?

7. Prove that the representation of a permutation as the product of disjoint cycles is unique except possibly for the order of the factors.

8. The *order* of a permutation π is the least integer $p \, (> 0)$ such that $\pi^p = \epsilon$.
(a) Every permutation has an order.
(b) What is the order of a p-cycle?
(c) If σ is a p-cycle, τ is a q-cycle, and σ and τ are disjoint, what is the order of $\sigma\tau$?
(d) Give an example to show that the assumption of disjointness is essential in (c).
(e) If π is a permutation of order p and if $\pi^q = \epsilon$, then q is divisible by p.

9. Every permutation in S_k $(k > 1)$ can be written as a product, each factor of which is one of the transpositions $(1, 2), (1, 3), (1, 4), \cdots, (1, k)$.

10. Two permutations σ and τ are called *conjugate* if there exists a permutation π such that $\sigma\pi = \pi\tau$. Prove that σ and τ are conjugate if and only if they have the same cycle structure. (This means that in the representation of σ as a product of disjoint cycles, the number of p-cycles is, for each p, the same as the corresponding number for τ.)

§ 28. Parity

Since $(1, 3)(1, 2) = (1, 2)(2, 3)(= (1, 2, 3))$, we see that the representation of a permutation (even a cycle) as a product of transpositions is not necessarily unique. Since $(1, 3)(1, 4)(1, 2)(3, 4)(3, 2) = (1, 4)(1, 3)(1, 2)$ $(= (1, 2, 3, 4))$, we see that even the number of transpositions needed to factor a cycle is not necessarily unique. There is, nevertheless, something unique about the factorization, namely, whether the number of transpositions needed is even or odd. We proceed to state this result precisely, and to prove it.

Assume, for simplicity of notation, that $k = 4$. Let f be the polynomial (in four variables t_1, t_2, t_3, t_4) defined by

$$f(t_1, t_2, t_3, t_4) = (t_1 - t_2)(t_1 - t_3)(t_1 - t_4)(t_2 - t_3)(t_2 - t_4)(t_3 - t_4).$$

(In the general case f is the product of all the differences $t_i - t_j$ with $1 \leq i < j \leq k$.) Each permutation π in S_4 converts f into a new polynomial, denoted by πf; by definition

$$(\pi f)(t_1, t_2, t_3, t_4) = f(t_{\pi(1)}, t_{\pi(2)}, t_{\pi(3)}, t_{\pi(4)}).$$

In words: to obtain πf, replace each variable in f by the one whose subscript is obtained by allowing π to act on the subscript of the given one. If, for instance, $\tau = (2, 4)$, then

$$(\tau f)(t_1, t_2, t_3, t_4) = (t_1 - t_4)(t_1 - t_3)(t_1 - t_2)(t_4 - t_3)(t_4 - t_2)(t_3 - t_2).$$

If $\sigma = (1, 3, 2, 4)$, so that $\sigma\tau = (1, 3, 2)$, then both $(\sigma(\tau f))(t_1, t_2, t_3, t_4)$ and $((\sigma\tau)f)(t_1, t_2, t_3, t_4)$ are equal to

$$(t_3 - t_1)(t_3 - t_2)(t_3 - t_4)(t_1 - t_2)(t_1 - t_4)(t_2 - t_4).$$

These computations illustrate, and indicate the proofs of, three important facts. (1) *For every permutation* π, *the factors of* πf *are the same as the factors of* f, *except possibly for sign and order*; consequently $\pi f = f$ or else $\pi f = -f$. The permutation π is called *even* if $\pi f = f$ and *odd* if $\pi f = -f$. The *signum* (or *sign*) of a permutation π, denoted by sgn π, is $+1$ or -1 according as π is even or odd, so that we always have $\pi f = (\text{sgn } \pi)f$. The fact that π is even, or odd, is sometimes expressed by saying that the *parity* of π is even, or odd, respectively. (2) *If* τ *is a transposition, then* sgn $\tau = -1$, *or, equivalently, every transposition is odd*. The proof is the obvious generalization of the following reasoning about the special example $(2, 4)$. Exactly one factor of f contains both t_2 and t_4, and that one changes sign in the passage from f to πf. If a factor contains neither t_2 nor t_4, it stays fixed. The factors containing only one of t_2 and t_4 come in pairs (such as the pair $(t_2 - t_3)$ and $(t_3 - t_4)$, or the pair $(t_1 - t_2)$ and $(t_1 - t_4)$). Each factor in such a pair goes into the other factor, except possibly that its sign may change; if it changes for one factor, it will change for its mate. (3) *If* σ *and* τ *are permutations, then* $(\sigma\tau)f = \sigma(\tau f)$; consequently $\sigma\tau$ is even if and only if σ and τ have the same parity. Observe that sgn $(\sigma\tau) = (\text{sgn } \sigma)(\text{sgn } \tau)$.

It follows from (2) and (3) that a product of a bunch of transpositions is even if and only if there are an even number of them, and it is odd otherwise. (Note, in particular, by looking at the proof of § 27, Theorem 2, that a p-cycle is even if and only if p is odd; in other words, if σ is a p-cycle, then sgn $\sigma = (-1)^{p+1}$.) Conclusion: no matter how a permutation π is factored into transpositions, the number of factors is always even (this is the case if π is even), or else it is always odd (this is the case if π is odd).

The product of two even permutations is even; the inverse of an even permutation is even; the identity permutation is even. These facts are summed up by saying that the set of all even permutations is a *subgroup* of S_k; this subgroup (to be denoted by $\mathcal{C}_k$) is called the *alternating group* of *degree* k.

1. How many permutations are there in $\mathcal{Q}_k$?

2. Give examples of even permutations with even order and even permutations with odd order; do the same for odd permutations.

3. Every permutation in $\mathcal{Q}_k$ ($k > 2$) can be written as a product, each factor of which is one of the 3-cycles $(1, 2, 3)$, $(1, 2, 4)$, $\cdots$, $(1, 2, k)$.

§ 29. Multilinear forms

We are now ready to proceed with multilinear algebra. The basic concept is that of multilinear form (or functional), an easy generalization of the concept of bilinear form. Suppose that $\mathcal{V}_1, \cdots, \mathcal{V}_k$ are vector spaces (over the same field); a *k-linear* form ($k = 1, 2, 3, \cdots$) is a scalar-valued function on the direct sum $\mathcal{V}_1 \oplus \cdots \oplus \mathcal{V}_k$ with the property that for each fixed value of any $k - 1$ arguments it depends linearly on the remaining argument. The 1-linear forms are simply the linear functionals (on $\mathcal{V}_1$), and the 2-linear forms are the bilinear forms (on $\mathcal{V}_1 \oplus \mathcal{V}_2$). The 3-linear (or trilinear) forms are the scalar-valued functions w (on $\mathcal{V}_1 \oplus \mathcal{V}_2 \oplus \mathcal{V}_3$) such that

$$w(\alpha_1 x_1 + \alpha_2 x_2, y, z) = \alpha_1 w(x_1, y, z) + \alpha_2 w(x_2, y, z),$$

and such that similar identities hold for $w(x, \alpha_1 y_1 + \alpha_2 y_2, z)$ and $w(x, y, \alpha_1 z_1 + \alpha_2 z_2)$. A function that is k-linear for some k is called a *multilinear form*.

Much of the theory of bilinear forms extends easily to the multilinear case. Thus, for instance, if w_1 and w_2 are k-linear forms, if α_1 and α_2 are scalars, and if w is defined by

$$w(x_1, \cdots, x_k) = \alpha_1 w_1(x_1, \cdots, x_k) + \alpha_2 w_2(x_1, \cdots, x_k)$$

whenever x_i is in $\mathcal{V}_i$, $i = 1, \cdots, k$, then w is a k-linear form, denoted by $\alpha_1 w_1 + \alpha_2 w_2$. The set of all k-linear forms is a vector space with respect to this definition of the linear operations; the dimension of that vector space is the product $n_1 \cdots n_k$, where, of course, n_i is the dimension of $\mathcal{V}_i$. The proofs of all these statements are just like the proofs (in § 23) of the corresponding statements for the bilinear case. We could go on imitating the bilinear theory and, in particular, studying multiple tensor products. In order to hold our multilinear digression to a minimum, we shall proceed instead in a different, more special, and, for our purposes, more useful direction.

In what follows we shall restrict our attention to the case in which the k spaces $\mathcal{V}_i$ are all equal to one and the same vector space, say, $\mathcal{V}$; we shall assume that $\mathcal{V}$ is finite-dimensional. In this case we shall call a "k-linear form on $\mathcal{V}_1 \oplus \cdots \oplus \mathcal{V}_k$" simply a "$k$-linear form on $\mathcal{V}$," or, even more simply, a "k-linear form"; the language is slightly inaccurate but, in context, completely unambiguous. If the dimension of $\mathcal{V}$ is n, then the dimension of the vector space of all k-linear forms is n^k. The space $\mathcal{V}$ and, of course, the dimension n will be held fixed throughout the following discussion.

The special character of the case we are studying enables us to apply a technique that is not universally available; the technique is to operate on k-linear forms by permutations in S_k. If w is a k-linear form, and if π is in S_k, we write

$$\pi w(x_1, \cdots, x_k) = w(x_{\pi(1)}, \cdots, x_{\pi(k)})$$

whenever $x_1, \cdots, x_k$ are in $\mathcal{V}$. The function πw so defined is again a k-linear form. (The value of πw at $(x_1, \cdots, x_k)$ is more honestly denoted by $(\pi w)(x_1, \cdots, x_k)$; since, however, the simpler notation does not appear to lead to any confusion, we shall continue to use it.)

Using the way permutations act on k-linear forms, we can define some interesting sets of such forms. Thus, for instance, a k-linear form w is called *symmetric* if $\pi w = w$ for every permutation π in S_k. (Note that if $k = 1$, then this condition is trivially satisfied.) The set of all symmetric k-linear forms is a subspace of the space of all k-linear forms. Hence, in particular, the origin of that space, the k-linear form 0, is symmetric. For a non-trivial example, suppose that $k = 2$, let y_1 and y_2 be linear functionals on $\mathcal{V}$, and write

$$w(x_1, x_2) = y_1(x_1)y_2(x_2) + y_1(x_2)y_2(x_1).$$

This procedure for constructing k-linear forms has useful generalizations. Thus, for instance, if $1 \leq h < k \leq n$, and if u is an h-linear form and v is a $(k - h)$-linear form, then the equation

$$w(x_1, \cdots, x_k) = u(x_1, \cdots, x_h) \cdot v(x_{h+1}, \cdots, x_k)$$

defines a k-linear form w, which, in general, is not symmetric. A symmetric k-linear form can be obtained from w (or, for that matter, from any given k-linear form) by forming $\sum \pi w$, where the summation is extended over all permutations π in S_k.

We shall not study symmetric k-linear forms any more. We introduced them here because they constitute a very natural class of functions definable in terms of permutations. We abandon them now in favor of another class of functions, which play a much greater role in the theory.

§ 30. Alternating forms

A k-linear form w is *skew-symmetric* if $\pi w = -w$ for every odd permutation π in S_k. Equivalently, w is skew-symmetric if $\pi w = (\operatorname{sgn} \pi)w$ for every permutation π in S_k. (If $\pi w = (\operatorname{sgn} \pi)w$ for all π, then, in particular, $\pi w = -w$ whenever π is odd. If, conversely, $\pi w = -w$ for all odd π, then, given an arbitrary π, factor it into transpositions, say, $\pi = \tau_1 \cdots \tau_q$, observe that $\operatorname{sgn} \pi = (-1)^q$, and, since $\pi w = (-1)^q w$, conclude that $\pi w = (\operatorname{sgn} \pi)w$, as asserted. This proof makes tacit use of the unproved but easily available fact that if σ and τ are permutations in S_k, then $\sigma(\tau w) = (\sigma\tau)w$.) The set of all skew-symmetric k-linear forms is a subspace of the space of all k-linear forms. To get a non-trivial example of a skew-symmetric bilinear form w, let y_1 and y_2 be linear functionals and write

$$w(x_1, x_2) = y_1(x_1)y_2(x_2) - y_1(x_2)y_2(x_1).$$

More generally, if w is an arbitrary k-linear form, a skew-symmetric k-linear form can be obtained from w by forming $\sum (\operatorname{sgn} \pi)\pi w$, where the summation is extended over all permutations π in S_k.

A k-linear form w is called *alternating* if $w(x_1, \cdots, x_k) = 0$ whenever two of the x's are equal. (Note that if $k = 1$, then this condition is vacuously satisfied.) The set of all alternating k-linear forms is a subspace of the space of all k-linear forms. There is an important relation between alternating and skew-symmetric forms.

THEOREM 1. *Every alternating multilinear form is skew-symmetric.*

PROOF. Suppose that w is an alternating k-linear form, and that i and j are integers, $1 \leq i < j \leq k$. If $x_1, \cdots, x_k$ are vectors, we write

$$w_0(x_i, x_j) = w(x_1, \cdots, x_k);$$

if the x's other than x_i and x_j are held fixed (temporarily), then w_0 is an alternating bilinear form of its two arguments. Since, by bilinearity,

$$w_0(x_i + x_j, x_i + x_j) = w_0(x_i, x_i) + w_0(x_i, x_j) + w_0(x_j, x_i) + w_0(x_j, x_j),$$

and since, by the alternating character of w_0, the left side and the two extreme terms of the right side of this equation all vanish, we see that $w_0(x_j, x_i) = -w_0(x_i, x_j)$. This, however, says that

$$(i, j)w(x_1, \cdots, x_k) = -w(x_1, \cdots, x_k),$$

or, since the x's are arbitrary, that $(i, j)w = -w$. Since every odd permutation π is the product of an odd number of transpositions, such as (i, j), it follows that $\pi w = -w$ for every odd π, and the proof of the theorem is complete.

The connection between alternating forms and skew-symmetric ones involves one subtle point. Consider the following "proof" of the converse of Theorem 1: if w is a skew-symmetric k-linear form, if $1 \leq i < j \leq k$, and if $x_1, \cdots, x_k$ are vectors such that $x_i = x_j$, then $(i, j)w(x_1, \cdots, x_k) = w(x_1, \cdots, x_k)$ since $x_i = x_j$, and at the same time, $(i, j)w(x_1, \cdots, x_k) = -w(x_1, \cdots, x_k)$ since w is skew-symmetric; consequently $w(x_1, \cdots, x_k) = -w(x_1, \cdots, x_k)$, so that w is alternating. This argument is wrong; the trouble is in the inference "if $w = -w$, then $w = 0$." If we examine that inference in more detail, we find that it is based on the following reasoning: if $w = -w$, then $w + w = 0$, so that $(1 + 1)w = 0$. This is correct. The trouble is that in certain fields $1 + 1 = 0$, and therefore the inference from $(1 + 1)w = 0$ to $w = 0$ is not justified; the converse of Theorem 1 is, in fact, false for vector spaces over such fields.

THEOREM 2. *If $x_1, \cdots, x_k$ are linearly dependent vectors and if w is an alternating k-linear form, then $w(x_1, \cdots, x_k) = 0$.*

PROOF. If $x_i = 0$ for some i, the conclusion is trivial. If all the x_i are different from 0, we apply the theorem of § 6 to find an x_h, $2 \leq h \leq k$, that is a linear combination of the preceding ones. If, say, $x_h = \sum_{i=0}^{h-1} \alpha_i x_i$, replace x_h in $w(x_1, \cdots, x_k)$ by this expansion, use the linearity of $w(x_1, \cdots, x_k)$ in its h-th argument, and draw the desired conclusion by an $(h - 1)$-fold application of the assumption that w is alternating.

In one extreme case (namely, when $k = n$) a sort of converse of Theorem 2 is true.

THEOREM 3. *If w is a non-zero alternating n-linear form, and if $x_1, \cdots, x_n$ are linearly independent vectors, then $w(x_1, \cdots, x_n) \neq 0$.*

PROOF. Since (§ 8, Theorem 2) the vectors $x_1, \cdots, x_n$ form a basis, we may, given an arbitrary set of n vectors $y_1, \cdots, y_n$, write each y as a linear combination of the x's. If we replace each y in $w(y_1, \cdots, y_n)$ by the corresponding linear combination of x's and expand the result by multilinearity, we obtain a long linear combination of terms such as $w(z_1, \cdots, z_n)$, where each z is one of the x's. If, in such a term, two of the z's coincide, then, since w is alternating, that term must vanish. If, on the other hand, all the z's are distinct, then $w(z_1, \cdots, z_n) = \pi w(x_1, \cdots, x_n)$ for some permutation π. Since (Theorem 1) w is skew-symmetric, it follows that $w(z_1, \cdots, z_n) = (\text{sgn } \pi)w(x_1, \cdots, x_n)$. If $w(x_1, \cdots, x_n)$ were 0, it would follow that $w(z_1, \cdots, z_n) = 0$, and hence that $w(y_1, \cdots, y_n) = 0$ for all $y_1, \cdots, y_n$, contradicting the assumption that $w \neq 0$.

The proof (not the statement) of this result yields a valuable corollary.

Tʜᴇᴏʀᴇᴍ 4. *Any two alternating n-linear forms are linearly dependent.*

ᴘʀᴏᴏꜰ. Suppose that w_1 and w_2 are alternating n-linear forms and that $\{x_1, \cdots, x_n\}$ is a basis. Given any n vectors $y_1, \cdots, y_n$, write each of them as a linear combination of the x's, and, just as above, replace each of them, in both $w_1(y_1, \cdots, y_n)$ and $w_2(y_1, \cdots, y_n)$, by the corresponding linear combination. It follows that each of $w_1(y_1, \cdots, y_n)$ and $w_2(y_1, \cdots, y_n)$ is a linear combination (the same linear combination) of terms such as $w_1(z_1, \cdots, z_n)$ and $w_2(z_1, \cdots, z_n)$, where each z is one of the x's. Since $w_1(x_1, \cdots, x_n)$ and $w_2(x_1, \cdots, x_n)$ are scalars, they are linearly dependent, so that there exist scalars α_1 and α_2 not both zero, such that $\alpha_1 w_1(x_1, \cdots, x_n) + \alpha_2 w_2(x_1, \cdots, x_n) = 0$; from these facts we may infer that $\alpha_1 w_1 + \alpha_2 w_2 = 0$, as asserted.

§ 31. Alternating forms of maximal degree

Glancing back at the last section, the reader will observe that we did not give any non-trivial examples of alternating k-linear forms, and we did not even indirectly hint at any existence theorem concerning them. In fact they do not always exist; § 30, Theorem 2 implies, for instance, that if $k > n$, then 0 is the only alternating k-linear form. (See § 8, Theorem 2.) For the applications we have in mind, we need only one existence theorem; we proceed to prove a rather sharp form of it.

Tʜᴇᴏʀᴇᴍ. *If* n > 0, *the vector space of alternating n-linear forms on an n-dimensional vector space is one-dimensional.*

ᴘʀᴏᴏꜰ. We show first that if $1 \leq k \leq n$, then there exists at least one non-zero alternating k-linear form; the proof goes by induction on k. If $k = 1$, the desired result follows from the existence of non-trivial linear functionals (see § 15, Theorem 3). If $1 \leq k < n$, we assume that v is a non-zero alternating k-linear form; using v we shall construct a non-zero alternating $(k + 1)$-linear form w. Since $v \neq 0$, we can find vectors $x_1^0, \cdots, x_k^0$ such that $v(x_1^0, \cdots, x_k^0) \neq 0$ (the superscripts are just indices here). Since $k < n$, we can find a vector x_{k+1}^0 that does not belong to the subspace spanned by $x_1^0, \cdots, x_k^0$, and (see § 17, Theorem 1) then we can find a linear functional u such that $u(x_1^0) = \cdots = u(x_k^0) = 0$ and $u(x_{k+1}^0) \neq 0$.

The promised $(k + 1)$-linear form w is obtained from the linear functional u and the k-linear form v by writing

$$(1) \quad w(x_1, \cdots, x_k, x_{k+1}) = \sum_{i=1}^{k} (i, k+1)v(x_1, \cdots, x_k)u(x_{k+1})$$
$$- v(x_1, \cdots, x_k)u(x_{k+1}).$$

Thus, for instance, if $k = 3$, then

$$w(x_1, x_2, x_3, x_4) = v(x_4, x_2, x_3)u(x_1) + v(x_1, x_4, x_3)u(x_2)$$

$$+ v(x_1, x_2, x_4)u(x_3) - v(x_1, x_2, x_3)u(x_4).$$

It follows from the elementary discussion in § 29 that w is indeed a $(k + 1)$-linear form; we are to prove that it is non-zero and alternating.

The fact that w is not identically zero is easy to prove. Indeed, since $u(x_i^0) = 0$ for $i = 1, \cdots, k$, it follows that if we replace each x_i by x_i^0 in (1), $i = 1, \cdots, k + 1$, then the first k terms of the sum on the right all vanish, and, consequently,

$$(2) \qquad w(x_1^0, \cdots, x_k^0, x_{k+1}^0) = -v(x_1^0, \cdots, x_k^0)u(x_{k+1}^0) \neq 0.$$

Suppose now that $x_1, \cdots, x_k, x_{k+1}$ are vectors and i and j are integers such that $1 \leq i < j \leq k + 1$ and $x_i = x_j$. We are to prove that, under these circumstances, $w(x_1, \cdots, x_k, x_{k+1}) = 0$. We note that both x_i and x_j occur in the argument of v in all but two of the $k + 1$ terms on the right side of (1). Since v is alternating, the terms in which both x_i and x_j do so occur all vanish.

The remainder of the proof splits naturally into two cases. If $j = k + 1$, then all that is left is

$$(i, k + 1)v(x_1, \cdots, x_k)u(x_{k+1}) - v(x_1, \cdots, x_k)u(x_{k+1}),$$

and, since $x_i = x_{k+1}$, this is clearly equal to 0. If $j \leq k$, then each of the two possibly non-vanishing terms that are still left can be obtained from the other by an application of the transposition (i, j). It follows that those terms differ in sign only, and hence that their sum is zero. This proves that w is alternating and proves, therefore, that the dimension of the space of alternating n-linear forms is not less than 1.

The fact that the dimension of the space of alternating n-linear forms is not more than 1 is an immediate consequence of § 30, Theorem 4.

This concludes our discussion of multilinear algebra. The reader might well charge that the discussion was not very strongly motivated. The complete motivation cannot be contained in this book; the justification for studying multilinear algebra is the wide applicability of the subject. The only application that we shall make is to the theory of determinants (which, to be sure, could be treated by more direct but less elegant methods, involving much greater dependence on arbitrary choices of bases); that application belongs to the next chapter.

1. Interpret the following matrices as linear transformations on $\mathbb{C}^2$ or $\mathbb{C}^3$ and, in each case find a basis such that the matrix of the transformation with respect to that basis is triangular.

2. Give an example of a skew-symmetric multilinear form that is not alternating. (Recall that in view of the discussion in § 30 the field of scalars must have characteristic 2.)

3. Give an example of a non-zero alternating k-linear form w on an n-dimensional space ($k < n$), such that $w(x_1, \cdots, x_k) = 0$ for some set of linearly independent vectors $x_1, \cdots, x_k$.

4. What is the dimension of the space of all symmetric k-linear forms? What about the skew-symmetric ones? What about the alternating ones?

CHAPTER II

TRANSFORMATIONS

§ 32. Linear transformations

We come now to the objects that really make vector spaces interesting.

DEFINITION. A *linear transformation* (or *operator*) A on a vector space $\mathcal{V}$ is a correspondence that assigns to every vector x in $\mathcal{V}$ a vector Ax in $\mathcal{V}$, in such a way that

$$A(\alpha x + \beta y) = \alpha Ax + \beta Ay$$

identically in the vectors x and y and the scalars α and β.

We make again the remark that we made in connection with the definition of linear functionals, namely, that for a linear transformation A, as we defined it, $A0 = 0$. For this reason such transformations are sometimes called *homogeneous* linear transformations.

Before discussing any properties of linear transformations we give several examples. We shall not bother to prove that the transformations we mention are indeed linear; in all cases the verification of the equation that defines linearity is a simple exercise.

(1) Two special transformations of considerable importance for the study that follows, and for which we shall consistently reserve the symbols 0 and 1 respectively, are defined (for all x) by $0x = 0$ and $1x = x$.

(2) Let x_0 be any fixed vector in $\mathcal{V}$, and let y_0 be any linear functional on $\mathcal{V}$; write $Ax = y_0(x) \cdot x_0$. More generally: let $\{x_1, \cdots, x_n\}$ be an arbitrary finite set of vectors in $\mathcal{V}$ and let $\{y_1, \cdots, y_n\}$ be a corresponding set of linear functionals on $\mathcal{V}$; write $Ax = y_1(x)x_1 + \cdots + y_n(x)x_n$. It is not difficult to prove that if, in particular, $\mathcal{V}$ is n-dimensional, and the vectors $x_1, \cdots, x_n$ form a basis for $\mathcal{V}$, then every linear transformation A has the form just described.

(3) Let π be a permutation of the integers $\{1, \cdots, n\}$; if $x = (\xi_1, \cdots, \xi_n)$ is a vector in $\mathbb{C}^n$, write $Ax = (\xi_{\pi(1)}, \cdots, \xi_{\pi(n)})$. Similarly, let π be a polynomial with complex coefficients; if x is a vector (polynomial) in $\mathcal{P}$, write $Ax = y$ for the polynomial defined by $y(t) = x(\pi(t))$.

(4) For any x in $\mathcal{P}_n$, $x(t) = \sum_{j=0}^{n-1} \xi_j t^j$, write $(Dx)(t) = \sum_{j=0}^{n-1} j\xi_j t^{j-1}$. (We use the letter D here as a reminder that Dx is the derivative of the polynomial x. We remark that we might have defined D on $\mathcal{P}$ as well as on $\mathcal{P}_n$; we shall make use of this fact later. Observe that for polynomials the definition of differentiation can be given purely algebraically, and does not need the usual theory of limiting processes.)

(5) For every x in $\mathcal{P}$, $x(t) = \sum_{j=0}^{n-1} \xi_j t^j$, write $Sx = \sum_{j=0}^{n-1} \frac{\xi_j}{j+1} t^{j+1}$. (Once more we are disguising by algebraic notation a well-known analytic concept. Just as in (4) $(Dx)(t)$ stood for $\frac{dx}{dt}$, so here $(Sx)(t)$ is the same as $\int_0^t x(s)\, ds$.)

(6) Let m be a polynomial with complex coefficients in a variable t. (We may, although it is not particularly profitable to do so, consider m as an element of $\mathcal{P}$.) For every x in $\mathcal{P}$, we write Mx for the polynomial defined by $(Mx)(t) = m(t)x(t)$. For later purposes we introduce a special symbol; in case $m(t) = t$, we shall write T for the transformation M, so that $(Tx)(t) = tx(t)$.

§ 33. Transformations as vectors

We proceed now to derive certain elementary properties of, and relations among, linear transformations on a vector space. More particularly, we shall indicate several ways of making new transformations out of old ones; we shall generally be satisfied with giving the definition of the new transformations and we shall omit the proof of linearity.

If A and B are linear transformations, we define their *sum*, $S = A + B$, by the equation $Sx = Ax + Bx$ (for every x). We observe that the commutativity and associativity of addition in $\mathcal{V}$ imply immediately that the addition of linear transformations is commutative and associative. Much more than this is true. If we consider the sum of any linear transformation A and the linear transformation 0 (defined in the preceding section), we see that $A + 0 = A$. If, for each A, we denote by $-A$ the transformation defined by $(-A)x = -(Ax)$, we see that $A + (-A) = 0$, and that the transformation $-A$, so defined, is the only linear transformation B with the property that $A + B = 0$. To sum up: the properties of a vector space, described in the axioms (A) of § 2, appear again in the set of all linear transformations on the space; the set of all linear transformations is an abelian group with respect to the operation of addition.

We continue in the same spirit. By now it will not surprise anybody if the axioms (B) and (C) of vector spaces are also satisfied by the set of all linear transformations. They are. For any A, and any scalar α, we define the product αA by the equation $(\alpha A)x = \alpha(Ax)$. Axioms (B) and (C) are immediately verified; we sum up as follows.

THEOREM. *The set of all linear transformations on a vector space is itself a vector space.*

We shall usually ignore this theorem; the reason is that we can say much more about linear transformations, and the mere fact that they form a vector space is used only very rarely. The "much more" that we can say is that there exists for linear transformations a more or less decent definition of multiplication, which we discuss in the next section.

1. Prove that each of the correspondences described below is a linear transformation.

(a) $\mathcal{V}$ is the set $\mathcal{C}$ of complex numbers regarded as a real vector space; Ax is the complex conjugate of x.

(b) $\mathcal{V}$ is $\mathcal{P}$; if x is a polynomial, then $(Ax)(t) = x(t + 1) - x(t)$.

(c) $\mathcal{V}$ is the k-fold tensor product of a vector space with itself; A is such that $A(x_1 \otimes \cdots \otimes x_k) = x_{\pi(1)} \otimes \cdots \otimes x_{\pi(k)}$, where π is a permutation of $\{1, \cdots, k\}$.

(d) $\mathcal{V}$ is the set of all k-linear forms on a vector space; $(Aw)(x_1, \cdots, x_k) = w(x_{\pi(1)}, \cdots, x_{\pi(k)})$, where π is a permutation of $\{1, \cdots, k\}$.

(e) $\mathcal{V}$ is the set of all k-linear forms on a vector space; if w is in $\mathcal{V}$, then $Aw = \sum \pi w$, where the summation is extended over all permutations π in S_k.

(f) Same as (e) except that $Aw = \sum (\operatorname{sgn} \pi) \pi w$.

2. Prove that if $\mathcal{V}$ is a finite-dimensional vector space, then the space of all linear transformations on $\mathcal{V}$ is finite-dimensional, and find its dimension.

3. The concept of a "linear transformation," as defined in the text, is too special for some purposes. According to a more general definition, a linear transformation from a vector space $\mathcal{U}$ to a vector space $\mathcal{V}$ over the same field is a correspondence A that assigns to every vector x in $\mathcal{U}$ a vector Ax in $\mathcal{V}$ so that

$$A(\alpha x + \beta y) = \alpha Ax + \beta Ay.$$

Prove that each of the correspondences described below is a linear transformation in this generalized sense.

(a) $\mathcal{V}$ is the field of scalars of $\mathcal{U}$; A is a linear functional on $\mathcal{U}$.

(b) $\mathcal{U}$ is the direct sum of $\mathcal{V}$ with some other space; A maps each pair in $\mathcal{U}$ onto its first coordinate.

(c) $\mathcal{V}$ is the quotient of $\mathcal{U}$ modulo a subspace; A maps each vector in $\mathcal{U}$ onto the coset it determines.

(d) Let w be a bilinear functional on a direct sum $\mathfrak{U} \oplus \mathfrak{V}_0$. Let $\mathfrak{V}$ be the dual of $\mathfrak{V}_0$, and define A to be the correspondence that assigns to each x_0 in $\mathfrak{U}$ the linear functional on $\mathfrak{V}_0$ obtained from w by setting its first argument equal to x_0.

4. (a) Suppose that $\mathfrak{U}$ and $\mathfrak{V}$ are vector spaces over the same field. If A and B are linear transformations from $\mathfrak{U}$ to $\mathfrak{V}$, if α and β are scalars, and if

$$Cx = \alpha Ax + \beta Bx$$

for each x in $\mathfrak{U}$, then C is a linear transformation from $\mathfrak{U}$ to $\mathfrak{V}$.

(b) If we write, by definition, $C = \alpha A + \beta B$, then the set of all linear transformations from $\mathfrak{U}$ to $\mathfrak{V}$ becomes a vector space with respect to this definition of the linear operations.

(c) Prove that if $\mathfrak{U}$ and $\mathfrak{V}$ are finite-dimensional, then so is the space of all linear transformations from $\mathfrak{U}$ to $\mathfrak{V}$, and find its dimension.

5. Suppose that $\mathfrak{M}$ is an m-dimensional subspace of an n-dimensional vector space $\mathfrak{V}$. Prove that the set of those linear transformations A on $\mathfrak{V}$ for which $Ax = 0$ whenever x is in $\mathfrak{M}$ is a subspace of the set of all linear transformations on $\mathfrak{V}$, and find the dimension of that subspace.

§ 34. Products

The *product* P of two linear transformations A and B, $P = AB$, is defined by the equation $Px = A(Bx)$.

The notion of multiplication is fundamental for all that follows. Before giving any examples to illustrate the meaning of transformation products, let us observe the implications of the symbolism, $P = AB$. To say that P is a transformation means, of course, that given a vector x, P does something to it. What it does is found out by operating on x with B, that is, finding Bx, and then operating on the result with A. In other words, if we look on the symbol for a transformation as a recipe for performing a certain act, then the symbol for the product of two transformations is to be read from right to left. The order to transform by AB means to transform first by B and then by A. This may seem like an undue amount of fuss to raise about a small point; however, as we shall soon see, transformation multiplication is, in general, not commutative, and the order in which we transform makes a lot of difference.

The most notorious example of non-commutativity is found on the space $\mathcal{O}$. We consider the differentiation and multiplication transformations D and T, defined by $(Dx)(t) = \dfrac{dx}{dt}$ and $(Tx)(t) = tx(t)$; we have

$$(DTx)(t) = \frac{d}{dt}(tx(t)) = x(t) + t\frac{dx}{dt}$$

and

$$(TDx)(t) = t\frac{dx}{dt}.$$

In other words, not only is it false that $DT = TD$ (so that $DT - TD = 0$), but, in fact, $(DT - TD)x = x$ for every x, so that $DT - TD = 1$.

On the basis of the examples in § 32, the reader should be able to construct many examples of pairs of non-commutative transformations. Those who are used to thinking of linear transformations geometrically can, for example, readily convince themselves that the product of two rotations of $\mathfrak{R}^3$ (about the origin) depends in general on the order in which they are performed.

Most of the formal algebraic properties of numerical multiplication (with the already mentioned notable exception of commutativity) are valid in the algebra of transformations. Thus we have

(1) $$A0 = 0A = 0,$$

(2) $$A1 = 1A = A,$$

(3) $$A(B + C) = AB + AC,$$

(4) $$(A + B)C = AC + BC,$$

(5) $$A(BC) = (AB)C.$$

The proofs of all these identities are immediate consequences of the definitions of addition and multiplication; to illustrate the principle we prove (3), one of the distributive laws. The proof consists of the following computation:

$$(A(B + C))x = A((B + C)x) = A(Bx + Cx)$$

$$= A(Bx) + A(Cx) = (AB)x + (AC)x$$

$$= (AB + AC)x.$$

§ 35. Polynomials

The associative law of multiplication enables us to write the product of three (or more) factors without any parentheses; in particular we may consider the product of any finite number, say, m, of factors all equal to A. This product depends only on A and on m (and not, as we just remarked, on any bracketing of the factors); we shall denote it by A^m. The justification for this notation is that, although in general transformation multiplication is not commutative, for the powers of one transformation we do have the usual laws of exponents, $A^n A^m = A^{n+m}$ and $(A^n)^m = A^{nm}$. We observe that $A^1 = A$; it is customary also to write, by definition, $A^0 = 1$. With these definitions the calculus of powers of a single transformation is almost exactly the same as in ordinary arithmetic. We may, in particular, define polynomials in a linear transformation. Thus if p is any polynomial with scalar coefficients in a variable t, say $p(t) = \alpha_0 + \alpha_1 t + \cdots + \alpha_n t^n$, we may form the linear transformation

$$p(A) = \alpha_0 1 + \alpha_1 A + \cdots + \alpha_n A^n.$$

The rules for the algebraic manipulation of such polynomials are easy. Thus $p(t)q(t) = r(t)$ implies $p(A)q(A) = r(A)$ (so that, in particular, any $p(A)$ and $q(A)$ are commutative); if $p(t) = \alpha$ (identically), we shall usually write $p(A) = \alpha$ (instead of $p(A) = \alpha \cdot 1$); this is in harmony with the use of the symbols 0 and 1 for linear transformations.

If p is a polynomial in two variables and if A and B are linear transformations, it is not usually possible to give any sensible interpretation to $p(A, B)$. The trouble, of course, is that A and B may not commute, and even a simple monomial, such as $s^2 t$, will cause confusion. If $p(s, t) = s^2 t$, what should we mean by $p(A, B)$? Should it be $A^2 B$, or ABA, or BA^2? It is important to recognize that there is a difficulty here; fortunately for us it is not necessary to try to get around it. We shall work with polynomials in several variables only in connection with commutative transformations, and then everything is simple. We observe that if $AB = BA$, then $A^n B^m = B^m A^n$, and therefore $p(A, B)$ has an unambiguous meaning for every polynomial p. The formal properties of the correspondence between (commutative) transformations and polynomials are just as valid for several variables as for one; we omit the details.

For an example of the possible behavior of the powers of a transformation we look at the differentiation transformation D on $\mathcal{P}$ (or, just as well, on $\mathcal{P}_n$, for some n). It is easy to see that for every positive integer k, and for every polynomial x in $\mathcal{P}$, we have $(D^k x)(t) = \dfrac{d^k x}{dt^k}$. We observe that whatever else D does, it lowers the degree of the polynomial on which it acts by exactly one unit (assuming, of course, that the degree is $\geqq 1$). Let x be a polynomial of degree $n - 1$, say; what is $D^n x$? Or put it another way: what is the product of the two (commutative) transformations D^k and D^{n-k} (where k is any integer between 0 and n), considered on the space $\mathcal{P}_n$? We mention this example to bring out the disconcerting fact implied by the answer to the last question; the product of two transformations may vanish even though neither one of them is zero. A non-zero transformation whose product with some non-zero transformation is zero is called a *divisor of zero*.

1. Calculate the linear transformations D^nS^n and S^nD^n, $n = 1, 2, 3, \cdots$; in other words, compute the effect of each such transformation on an arbitrary element of $\mathcal{P}$. (Here D and S denote the differentiation and integration transformations defined in § 32.)

2. If A and B are linear transformations such that $AB - BA$ commutes with A, then $A^kB - BA^k = kA^{k-1}(AB - BA)$ for every positive integer k.

3. Suppose that $Ax(t) = x(t + 1)$ for every x in $\mathcal{P}_n$; prove that if D is the differentiation operator, then

$$1 + \frac{D}{1!} + \frac{D^2}{2!} + \cdots + \frac{D^{n-1}}{(n-1)!} = A.$$

4. (a) If A is a linear transformation on an n-dimensional vector space, then there exists a non-zero polynomial p of degree $\leq n^2$ such that $p(A) = 0$.
(b) If $Ax = y_0(x)x_0$ (see § 32, (2)), find a non-zero polynomial p such that $p(A) = 0$. What is the smallest possible degree p can have?

5. The product of linear transformations between different vector spaces is defined only if they "match" in the following sense. Suppose that $\mathcal{U}$, $\mathcal{V}$, and $\mathcal{W}$ are vector spaces over the same field, and suppose that A and B are linear transformations from $\mathcal{U}$ to $\mathcal{V}$ and from $\mathcal{V}$ to $\mathcal{W}$, respectively. The product $C = BA$ (the order is important) is defined to be the linear transformation from $\mathcal{U}$ to $\mathcal{W}$ given by $Cx = B(Ax)$. Interpret and prove as many as possible among the equations § 34, (1)–(5) for this concept of multiplication.

6. Let A be a linear transformation on an n-dimensional vector space $\mathcal{V}$.
(a) Prove that the set of all those linear transformations B on $\mathcal{V}$ for which $AB = 0$ is a subspace of the space of all linear transformations on $\mathcal{V}$.
(b) Show that by a suitable choice for A the dimension of the subspace described in (a) can be made to equal 0, or n, or n^2. What values can this dimension attain?
(c) Can every subspace of the space of all linear transformations be obtained in the manner described in (a) (by the choice of a suitable A)?

7. Let A be a linear transformation on a vector space $\mathcal{V}$, and consider the correspondence that assigns to each linear transformation X on $\mathcal{V}$ the linear transformation AX. Prove that this correspondence is a linear transformation (on the space of all linear transformations). Can every linear transformation on that space be obtained in this manner (by the choice of a suitable A)?

§ 36. Inverses

In each of the two preceding sections we gave an example; these two examples bring out the two nasty properties that the multiplication of linear transformations has, namely, non-commutativity and the existence of divisors of zero. We turn now to the more pleasant properties that linear transformations sometimes have.

It may happen that the linear transformation A has one or both of the following two very special properties.

(i) If $x_1 \neq x_2$, then $Ax_1 \neq Ax_2$.

(ii) To every vector y there corresponds (at least) one vector x such that $Ax = y$.

If ever A has both these properties we shall say that A is *invertible*. If A is invertible, we define a linear transformation, called the *inverse* of A and denoted by A^{-1}, as follows. If y_0 is any vector, we may (by (ii)) find an x_0 for which $Ax_0 = y_0$. This x_0 is, moreover, uniquely determined, since $x_0 \neq x_1$ implies (by (i)) that $y_0 = Ax_0 \neq Ax_1$. We define $A^{-1}y_0$ to be x_0. To prove that A^{-1} is linear, we evaluate $A^{-1}(\alpha_1 y_1 + \alpha_2 y_2)$. If $Ax_1 = y_1$ and $Ax_2 = y_2$, then the linearity of A tells us that $A(\alpha_1 x_1 + \alpha_2 x_2) = \alpha_1 y_1 + \alpha_2 y_2$, so that $A^{-1}(\alpha_1 y_1 + \alpha_2 y_2) = \alpha_1 x_1 + \alpha_2 x_2 = \alpha_1 A^{-1}y_1 + \alpha_2 A^{-1}y_2$.

As a trivial example of an invertible transformation we mention the identity transformation 1; clearly $1^{-1} = 1$. The transformation 0 is not invertible; it violates both the conditions (i) and (ii) about as strongly as they can be violated.

It is immediate from the definition that for any invertible A we have

$$AA^{-1} = A^{-1}A = 1;$$

we shall now show that these equations serve to characterize A^{-1}.

Tʜᴇᴏʀᴇᴍ 1. *If A, B, and C are linear transformations such that*

$$AB = CA = 1,$$

then A is invertible and $A^{-1} = B = C$.

ᴘʀᴏᴏꜰ. If $Ax_1 = Ax_2$, then $CAx_1 = CAx_2$, so that (since $CA = 1$) $x_1 = x_2$; in other words, the first condition of the definition of invertibility is satisfied. The second condition is also satisfied, for if y is any vector and $x = By$, then $y = ABy = Ax$. Multiplying $AB = 1$ on the left, and $CA = 1$ on the right, by A^{-1}, we see that $A^{-1} = B = C$.

To show that neither $AB = 1$ nor $CA = 1$ is, by itself, sufficient to ensure the invertibility of A, we call attention to the differentiation and integration transformations D and S, defined in § 32, (4) and (5). Although $DS = 1$, neither D nor S is invertible; D violates (i), and S violates (ii).

In finite-dimensional spaces the situation is much simpler.

Tʜᴇᴏʀᴇᴍ 2. *A linear transformation A on a finite-dimensional vector space $\mathcal{V}$ is invertible if and only if $Ax = 0$ implies that $x = 0$, or, alternatively, if and only if every y in $\mathcal{V}$ can be written in the form $y = Ax$.*

PROOF. If A is invertible, both conditions are satisfied; this much is trivial. Suppose now that $Ax = 0$ implies that $x = 0$. Then $u \neq v$, that is, $u - v \neq 0$, implies that $A(u - v) \neq 0$, that is, that $Au \neq Av$; this proves (i). To prove (ii), let $\{x_1, \cdots, x_n\}$ be a basis in $\mathcal{U}$; we assert that $\{Ax_1, \cdots, Ax_n\}$ is also a basis. According to § 8, Theorem 2, we need only prove linear independence. But $\sum_i \alpha_i Ax_i = 0$ means $A(\sum_i \alpha_i x_i) = 0$, and, by hypothesis, this implies that $\sum_i \alpha_i x_i = 0$; the linear independence of the x_i now tells us that $\alpha_1 = \cdots = \alpha_n = 0$. It follows, of course, that every vector y may be written in the form $y = \sum_i \alpha_i Ax_i = A(\sum_i \alpha_i x_i)$.

Let us assume next that every y is an Ax, and let $\{y_1, \cdots, y_n\}$ be any basis in $\mathcal{U}$. Corresponding to each y_i we may find a (not necessarily unique) x_i for which $y_i = Ax_i$; we assert that $\{x_1, \cdots, x_n\}$ is also a basis. For $\sum_i \alpha_i x_i = 0$ implies $\sum_i \alpha_i Ax_i = \sum_i \alpha_i y_i = 0$, so that $\alpha_1 = \cdots = \alpha_n = 0$. Consequently every x may be written in the form $x = \sum_i \alpha_i x_i$, and $Ax = 0$ implies, as in the argument just given, that $x = 0$.

THEOREM 3. *If A and B are invertible, then AB is invertible and $(AB)^{-1} = B^{-1}A^{-1}$. If A is invertible and $\alpha \neq 0$, then αA is invertible and $(\alpha A)^{-1} = \frac{1}{\alpha}A^{-1}$. If A is invertible, then A^{-1} is invertible and $(A^{-1})^{-1} = A$.*

PROOF. According to Theorem 1, it is sufficient to prove (for the first statement) that the product of AB with $B^{-1}A^{-1}$, in both orders, is the identity; this verification we leave to the reader. The proofs of both the remaining statements are identical in principle with this proof of the first statement; the last statement, for example, follows from the fact that the equations $AA^{-1} = A^{-1}A = 1$ are completely symmetric in A and A^{-1}.

We conclude our discussion of inverses with the following comment. In the spirit of the preceding section we may, if we like, define rational functions of A, whenever possible, by using A^{-1}. We shall not find it useful to do this, except in one case: if A is invertible, then we know that A^n is also invertible, $n = 1, 2, \cdots$; we shall write A^{-n} for $(A^n)^{-1}$, so that $A^{-n} = (A^{-1})^n$.

EXERCISES

1. Which of the linear transformations described in § 33, Ex. 1 are invertible?

2. A linear transformation A is defined on $\mathbb{C}^2$ by

$$A(\xi_1, \xi_2) = (\alpha\xi_1 + \beta\xi_2, \gamma\xi_1 + \delta\xi_2),$$

where $\alpha, \beta, \gamma,$ and δ are fixed scalars. Prove that A is invertible if and only if $\alpha\delta - \beta\gamma \neq 0$.

3. If A and B are linear transformations (on the same vector space), then a necessary and sufficient condition that both A and B be invertible is that both AB and BA be invertible.

4. If A and B are linear transformations on a finite-dimensional vector space, and if $AB = 1$, then both A and B are invertible.

5. (a) If A, B, C, and D are linear transformations (all on the same vector space), and if both $A + B$ and $A - B$ are invertible, then there exist linear transformations X and Y such that

$$AX + BY = C$$

and

$$BX + AY = D.$$

(b) To what extent are the invertibility assumptions in (a) necessary?

6. (a) A linear transformation on a finite-dimensional vector space is invertible if and only if it preserves linear independence. To say that A preserves linear independence means that whenever $\mathfrak{X}$ is a linearly independent set in the space $\mathcal{V}$ on which A acts, then $A\mathfrak{X}$ is also a linearly independent set in $\mathcal{V}$. (The symbol $A\mathfrak{X}$ denotes, of course, the set of all vectors of the form Ax, with x in $\mathfrak{X}$.)

(b) Is the assumption of finite-dimensionality needed for the validity of (a)?

7. Show that if A is a linear transformation such that $A^2 - A + 1 = 0$, then A is invertible.

8. If A and B are linear transformations (on the same vector space) and if $AB = 1$, then A is called a *left inverse* of B and B is called a *right inverse* of A. Prove that if A has exactly one right inverse, say B, then A is invertible. (Hint: consider $BA + B - 1$.)

9. If A is an invertible linear transformation on a finite-dimensional vector space $\mathcal{V}$, then there exists a polynomial p such that $A^{-1} = p(A)$. (Hint: find a non-zero polynomial q of least degree such that $q(A) = 0$ and prove that its constant term cannot be 0.)

10. Devise a sensible definition of invertibility for linear transformations from one vector space to another. Using that definition, decide which (if any) of the linear transformations described in § 33, Ex. 3 are invertible.

§ 37. Matrices

Let us now pick up the loose threads; having introduced the new concept of linear transformation, we must now find out what it has to do with the old concepts of bases, linear functionals, etc.

One of the most important tools in the study of linear transformations on finite-dimensional vector spaces is the concept of a matrix. Since this concept usually has no decent analogue in infinite-dimensional spaces, and since it is possible in most considerations to do without it, we shall try not to use it in proving theorems. It is, however, important to know what a matrix is; we enter now into the detailed discussion.

DEFINITION. Let $\mathcal{V}$ be an n-dimensional vector space, let $\mathcal{X} = \{x_1, \cdots, x_n\}$ be any basis of $\mathcal{V}$, and let A be a linear transformation on $\mathcal{V}$. Since every vector is a linear combination of the x_i, we have in particular

$$Ax_j = \sum_i \alpha_{ij} x_i$$

for $j = 1, \cdots, n$. The set (α_{ij}) of n^2 scalars, indexed with the double subscript i, j, is the *matrix* of A in the coordinate system $\mathcal{X}$; we shall generally denote it by $[A]$, or, if it becomes necessary to indicate the particular basis $\mathcal{X}$ under consideration, by $[A; \mathcal{X}]$. A matrix (α_{ij}) is usually written in the form of a square array:

$$[A] = \begin{bmatrix} \alpha_{11} & \alpha_{12} & \cdots & \alpha_{1n} \\ \alpha_{21} & \alpha_{22} & \cdots & \alpha_{2n} \\ \cdot & \cdot & & \cdot \\ \cdot & \cdot & & \cdot \\ \cdot & \cdot & & \cdot \\ \alpha_{n1} & \alpha_{n2} & \cdots & \alpha_{nn} \end{bmatrix};$$

the scalars $(\alpha_{i1}, \cdots, \alpha_{in})$ form a *row*, and $(\alpha_{1j}, \cdots, \alpha_{nj})$ a *column*, of $[A]$.

This definition does not define "matrix"; it defines "the matrix associated under certain conditions with a linear transformation." It is often useful to consider a matrix as something existing in its own right as a square array of scalars; in general, however, a matrix in this book will be tied up with a linear transformation and a basis.

We comment on notation. It is customary to use the same symbol, say, A, for the matrix as for the transformation. The justification for this is to be found in the discussion below (of properties of matrices). We do not follow this custom here, because one of our principal aims, in connection with matrices, is to emphasize that they depend on a coordinate system (whereas the notion of linear transformation does not), and to study how the relation between matrices and linear transformations changes as we pass from one coordinate system to another.

We call attention also to a peculiarity of the indexing of the elements α_{ij} of a matrix $[A]$. A basis is a basis, and so far, although we usually indexed its elements with the first n positive integers, the order of the elements in it was entirely immaterial. It is customary, however, when speaking of matrices, to refer to, say, the first row or the first column. This language is justified only if we think of the elements of the basis $\mathcal{X}$ as arranged in a definite order. Since in the majority of our considerations the order of the rows and the columns of a matrix is as irrelevant as the order of the elements of a basis, we did not include this aspect of matrices in our definition. It is important, however, to realize that the appearance of the square array associated with $[A]$ varies with the ordering of $\mathcal{X}$.

Everything we shall say about matrices can, accordingly, be interpreted from two different points of view; either in strict accordance with the letter of our definition, or else following a modified definition which makes correspond a matrix (with ordered rows and columns) not merely to a linear transformation and a basis, but also to an ordering of the basis.

One more word to those in the know. It is a perversity not of the author, but of nature, that makes us write

$$Ax_j = \sum_i \alpha_{ij} x_i,$$

instead of the more usual equation

$$Ax_i = \sum_j \alpha_{ij} x_j.$$

The reason is that we want the formulas for matrix multiplication and for the application of matrices to numerical vectors (that is, vectors $(\xi_1, \cdots, \xi_n)$ in $\mathbb{C}^n$) to appear normal, and somewhere in the process of passing from vectors to their coordinates the indices turn around. To state our rule explicitly: write Ax_j as a linear combination of $x_1, \cdots, x_n$, and write the coefficients so obtained as the j-th *column* of the matrix $[A]$. (The first index on α_{ij} is always the row index; the second one, the column index.)

For an example we consider the differentiation transformation D on the space $\mathcal{P}_n$, and the basis $\{x_1, \cdots, x_n\}$ defined by $x_i(t) = t^{i-1}$, $i = 1, \cdots, n$. What is the matrix of D in this basis? We have

$$
\begin{aligned}
Dx_1 &= 0x_1 + 0x_2 + \cdots + & 0x_{n-1} + 0x_n \\
Dx_2 &= 1x_1 + 0x_2 + \cdots + & 0x_{n-1} + 0x_n \\
Dx_3 &= 0x_1 + 2x_2 + \cdots + & 0x_{n-1} + 0x_n \\
& \quad \vdots \qquad\qquad \cdots \qquad\qquad \vdots \\
Dx_n &= 0x_1 + 0x_2 + \cdots + (n-1)x_{n-1} + 0x_n,
\end{aligned}
$$

(1)

so that

(2)
$$[D] = \begin{bmatrix} 0 & 1 & 0 & \cdots & 0 & 0 \\ 0 & 0 & 2 & \cdots & 0 & 0 \\ & \vdots & & & & \vdots \\ 0 & 0 & 0 & \cdots & 0 & n-1 \\ 0 & 0 & 0 & \cdots & 0 & 0 \end{bmatrix}.$$

The unpleasant phenomenon of indices turning around is seen by comparing (1) and (2).

§ 38. Matrices of transformations

There is now a certain amount of routine work to be done, most of which we shall leave to the imagination. The problem is this: in a fixed coordinate system $\mathfrak{X} = \{x_1, \cdots, x_n\}$, knowing the matrices of A and B, how can we find the matrices of $\alpha A + \beta B$, of AB, of 0, 1, etc.?

Write $[A] = (\alpha_{ij})$, $[B] = (\beta_{ij})$, $C = \alpha A + \beta B$, $[C] = (\gamma_{ij})$; we assert that

$$\gamma_{ij} = \alpha \alpha_{ij} + \beta \beta_{ij};$$

also if $[0] = (o_{ij})$ and $[1] = (e_{ij})$, then

$$o_{ij} = 0$$

and

$$e_{ij} = \delta_{ij} \; (= \text{the Kronecker delta}).$$

A more complicated rule is the following: if $C = AB$, $[C] = (\gamma_{ij})$, then

$$\gamma_{ij} = \sum_k \alpha_{ik}\beta_{kj}.$$

To prove this we use the definition of the matrix associated with a transformation, and juggle, thus:

$$Cx_j = A(Bx_j) = A(\sum_k \beta_{kj}x_k) = \sum_k \beta_{kj}Ax_k$$
$$= \sum_k \beta_{kj}(\sum_i \alpha_{ik}x_i) = \sum_i (\sum_k \alpha_{ik}\beta_{kj})x_i.$$

The relation between transformations and matrices is exactly the same as the relation between vectors and their coordinates, and the analogue of the isomorphism theorem of § 9 is true in the best possible sense. We shall make these statements precise.

With the aid of a fixed basis $\mathfrak{X}$, we have made correspond a matrix $[A]$ to every linear transformation A; the correspondence is described by the relations $Ax_j = \sum_i \alpha_{ij}x_i$. We assert now that this correspondence is one-to-one (that is, that the matrices of two different transformations are different), and that every array (α_{ij}) of n^2 scalars is the matrix of some transformation. To prove this, we observe in the first place that knowledge of the matrix of A completely determines A (that is, that Ax is thereby uniquely defined for every x), as follows: if $x = \sum_j \xi_j x_j$, then $Ax = \sum_j \xi_j Ax_j = \sum_j \xi_j(\sum_i \alpha_{ij}x_i) = \sum_i (\sum_j \alpha_{ij}\xi_j)x_i$. (In other words, if $y = Ax = \sum_i \eta_i x_i$, then

$$\eta_i = \sum_j \alpha_{ij}\xi_j.$$

Compare this with the comments in § 37 on the perversity of indices.) In the second place, there is no law against reading the relation $Ax_j = \sum_i \alpha_{ij}x_i$ backwards. If, in other words, (α_{ij}) is any array, we may use this relation to define a linear transformation A; it is clear that the matrix of A will be exactly (α_{ij}). (Once more, however, we emphasize the fundamental fact that this one-to-one correspondence between transformations and matrices was set up by means of a particular coordinate system, and that, as we pass from one coordinate system to another, the same linear transformation may correspond to several matrices, and one matrix may be the correspondent of many linear transformations.) The following statement sums up the essential part of the preceding discussion.

THEOREM. *Among the set of all matrices* (α_{ij}), (β_{ij}), *etc.*, $i, j = 1, \cdots, n$ *(not considered in relation to linear transformations), we define sum, scalar multiplication, product,* (o_{ij}), *and* (e_{ij}), *by*

$$(\alpha_{ij}) + (\beta_{ij}) = (\alpha_{ij} + \beta_{ij}),$$

$$\alpha(\alpha_{ij}) = (\alpha\alpha_{ij}),$$

$$(\alpha_{ij})(\beta_{ij}) = (\textstyle\sum_k \alpha_{ik}\beta_{kj}),$$

$$o_{ij} = 0, \quad e_{ij} = \delta_{ij}.$$

Then the correspondence (established by means of an arbitrary coordinate system $\mathfrak{X} = \{x_1, \cdots, x_n\}$ *of the n-dimensional vector space* $\mathcal{V}$), *between all linear transformations* A *on* $\mathcal{V}$ *and all matrices* (α_{ij}), *described by* $Ax_j = \sum_i \alpha_{ij}x_i$, *is an isomorphism; in other words, it is a one-to-one correspondence that preserves sum, scalar multiplication, product, 0, and 1.*

We have carefully avoided discussing the matrix of A^{-1}. It is possible to give an expression for $[A^{-1}]$ in terms of the elements α_{ij} of $[A]$, but the expression is not simple and, fortunately, not useful for us.

EXERCISES

1. Let A be the linear transformation on $\mathcal{P}_n$ defined by $(Ax)(t) = x(t + 1)$, and let $\{x_0, \cdots, x_{n-1}\}$ be the basis of $\mathcal{P}_n$ defined by $x_j(t) = t^j, j = 0, \cdots, n - 1$. Find the matrix of A with respect to this basis.

2. Find the matrix of the operation of conjugation on $\mathbb{C}$, considered as a real vector space, with respect to the basis $\{1, i\}$ (where $i = \sqrt{-1}$).

3. (a) Let π be a permutation of the integers $1, \cdots, n$; if $x = (\xi_1, \cdots, \xi_n)$ is a vector in $\mathbb{C}^n$, write $Ax = (\xi_{\pi(1)}, \cdots, \xi_{\pi(n)})$. If $x_i = (\delta_{i1}, \cdots, \delta_{in})$, find the matrix of A with respect to $\{x_1, \cdots, x_n\}$.
(b) Find all matrices that commute with the matrix of A.

4. Consider the vector space consisting of all real two-by-two matrices and let A be the linear transformation on this space that sends each matrix X onto PX, where $P = \begin{pmatrix} 1 & 1 \\ 1 & 1 \end{pmatrix}$. Find the matrix of A with respect to the basis consisting of $\begin{pmatrix} 1 & 0 \\ 0 & 0 \end{pmatrix}$, $\begin{pmatrix} 0 & 1 \\ 0 & 0 \end{pmatrix}$, $\begin{pmatrix} 0 & 0 \\ 1 & 0 \end{pmatrix}$, $\begin{pmatrix} 0 & 0 \\ 0 & 1 \end{pmatrix}$.

5. Consider the vector space consisting of all linear transformations on a vector space $\mathcal{V}$, and let A be the (left) multiplication transformation that sends each transformation X on $\mathcal{V}$ onto PX, where P is some prescribed transformation on $\mathcal{V}$. Under what conditions on P is A invertible?

6. Prove that if I, J, and K are the complex matrices

$$\begin{pmatrix} 0 & 1 \\ -1 & 0 \end{pmatrix}, \begin{pmatrix} 0 & i \\ i & 0 \end{pmatrix}, \begin{pmatrix} i & 0 \\ 0 & -i \end{pmatrix}$$

respectively (where $i = \sqrt{-1}$), then $I^2 = J^2 = K^2 = -1$, $IJ = -JI = K$, $JK = -KJ = I$, and $KI = -IK = J$.

7. (a) Prove that if A, B, and C are linear transformations on a two-dimensional vector space, then $(AB - BA)^2$ commutes with C.
(b) Is the conclusion of (a) true for higher-dimensional spaces?

8. Let A be the linear transformation on $\mathbb{C}^2$ defined by $A(\xi_1, \xi_2) = (\xi_1 + \xi_2, \xi_2)$. Prove that if a linear transformation B commutes with A, then there exists a polynomial p such that $B = p(A)$.

9. For which of the following polynomials p and matrices A is it true that $p(A) = 0$?

(a) $p(t) = t^3 - 3t^2 + 3t - 1$ $A = \begin{pmatrix} 1 & 1 & 1 \\ 0 & 1 & 1 \\ 0 & 0 & 1 \end{pmatrix}$.

(b) $p(t) = t^2 - 3t$, $A = \begin{pmatrix} 1 & 1 & 1 \\ 1 & 1 & 1 \\ 1 & 1 & 1 \end{pmatrix}$.

(c) $p(t) = t^3 + t^2 + t + 1$, $A = \begin{pmatrix} 1 & 1 & 0 \\ 1 & 1 & 1 \\ 0 & 1 & 1 \end{pmatrix}$.

(d) $p(t) = t^3 - 2t$, $A = \begin{pmatrix} 0 & 1 & 0 \\ 1 & 0 & 1 \\ 0 & 1 & 0 \end{pmatrix}$.

10. Prove that if A and B are the complex matrices

$$\begin{bmatrix} 0 & 1 & 0 & 0 \\ 0 & 0 & 1 & 0 \\ 0 & 0 & 0 & 1 \\ 1 & 0 & 0 & 0 \end{bmatrix} \text{ and } \begin{bmatrix} i & 0 & 0 & 0 \\ 0 & -1 & 0 & 0 \\ 0 & 0 & -i & 0 \\ 0 & 0 & 0 & 1 \end{bmatrix}$$

respectively (where $i = \sqrt{-1}$), and if $C = AB - iBA$, then $C^3 + C^2 + C = 0$.

11. If A and B are linear transformations on a vector space, and if $AB = 0$, does it follow that $BA = 0$?

12. What happens to the matrix of a linear transformation on a finite-dimensional vector space when the elements of the basis with respect to which the matrix is computed are permuted among themselves?

13. (a) Suppose that $\mathcal{U}$ is a finite-dimensional vector space with basis $\{x_1, \cdots, x_n\}$. Suppose that $\alpha_1, \cdots, \alpha_n$ are pairwise distinct scalars. If A is a linear transformation such that $Ax_j = \alpha_j x_j$, $j = 1, \cdots, n$, and if B is a linear transformation that commutes with A, then there exist scalars $\beta_1, \cdots, \beta_n$ such that $Bx_j = \beta_j x_j$.

(b) Prove that if B is a linear transformation on a finite-dimensional vector space $\mathcal{U}$ and if B commutes with every linear transformation on $\mathcal{U}$, then B is a scalar (that is, there exists a scalar β such that $Bx = \beta x$ for all x in $\mathcal{U}$).

14. If $\{x_1, \cdots, x_k\}$ and $\{y_1, \cdots, y_k\}$ are linearly independent sets of vectors in a finite-dimensional vector space $\mathcal{U}$, then there exists an invertible linear transformation A on $\mathcal{U}$ such that $Ax_j = y_j, j = 1, \cdots, k$.

15. If a matrix $[A] = (\alpha_{ij})$ is such that $\alpha_{ii} = 0$, $i = 1, \cdots, n$, then there exist matrices $[B] = (\beta_{ij})$ and $[C] = (\gamma_{ij})$ such that $[A] = [B][C] - [C][B]$. (Hint: try $\beta_{ij} = \beta_i \delta_{ij}$.)

16. Decide which of the following matrices are invertible and find the inverses of the ones that are.

(a) $\begin{pmatrix} 1 & 1 \\ 0 & 1 \end{pmatrix}$.

(b) $\begin{pmatrix} 1 & 1 \\ 1 & 1 \end{pmatrix}$.

(c) $\begin{pmatrix} 0 & 1 \\ 0 & 0 \end{pmatrix}$.

(d) $\begin{pmatrix} 0 & 1 \\ 1 & 0 \end{pmatrix}$.

(e) $\begin{pmatrix} 0 & 1 & 0 \\ 0 & 0 & 1 \\ 1 & 0 & 0 \end{pmatrix}$.

(f) $\begin{pmatrix} 1 & 0 & 1 \\ 1 & 0 & 1 \\ 1 & 0 & 1 \end{pmatrix}$.

g) $\begin{pmatrix} 0 & 1 & 0 \\ 1 & 0 & 1 \\ 0 & 1 & 0 \end{pmatrix}$.

17. For which values of α are the following matrices invertible? Find the inverses whenever possible.

(a) $\begin{pmatrix} \alpha & 1 \\ 1 & 0 \end{pmatrix}$.

(b) $\begin{pmatrix} 1 & \alpha \\ 1 & 0 \end{pmatrix}$.

(c) $\begin{pmatrix} 1 & \alpha \\ 1 & \alpha \end{pmatrix}$.

(d) $\begin{pmatrix} 1 & 1 \\ 1 & \alpha \end{pmatrix}$.

18. For which values of α are the following matrices invertible? Find the inverses whenever possible.

(a) $\begin{pmatrix} 1 & \alpha & 0 \\ \alpha & 1 & \alpha \\ 0 & \alpha & 1 \end{pmatrix}$.

(b) $\begin{pmatrix} \alpha & 1 & 0 \\ 1 & \alpha & 1 \\ 0 & 1 & \alpha \end{pmatrix}$.

(c) $\begin{pmatrix} 0 & 1 & \alpha \\ 1 & \alpha & 0 \\ \alpha & 0 & 1 \end{pmatrix}$.

(d) $\begin{pmatrix} 1 & 1 & 1 \\ 1 & 1 & \alpha \\ 1 & \alpha & 1 \end{pmatrix}$.

19. (a) It is easy to extend matrix theory to linear transformations between different vector spaces. Suppose that $\mathcal{U}$ and $\mathcal{U}$ are vector spaces over the same

field, let $\{x_1, \cdots, x_n\}$ and $\{y_1, \cdots, y_m\}$ be bases of $\mathfrak{U}$ and $\mathfrak{V}$ respectively, and let A be a linear transformation from $\mathfrak{U}$ to $\mathfrak{V}$. The matrix of A is, by definition, the rectangular, m by n, array of scalars defined by

$$Ax_j = \sum_i \alpha_{ij} y_i.$$

Define addition and multiplication of rectangular matrices so as to generalize as many as possible of the results of § 38. (Note that the product of an m_1 by n_1 matrix and an m_2 by n_2 matrix, in that order, will be defined only if $n_1 = m_2$.)

(b) Suppose that A and B are multipliable matrices. Partition A into four rectangular blocks (top left, top right, bottom left, bottom right) and then partition B similarly so that the number of columns in the top left part of A is the same as the number of rows in the top left part of B. If, in an obvious shorthand, these partitions are indicated by

$$A = \begin{pmatrix} A_{11} & A_{12} \\ A_{21} & A_{22} \end{pmatrix}, \quad B = \begin{pmatrix} B_{11} & B_{12} \\ B_{21} & B_{22} \end{pmatrix},$$

then

$$AB = \begin{pmatrix} A_{11}B_{11} + A_{12}B_{21} & A_{11}B_{12} + A_{12}B_{22} \\ A_{21}B_{11} + A_{22}B_{21} & A_{21}B_{12} + A_{22}B_{22} \end{pmatrix}.$$

(c) Use subspaces and complements to express the result of (b) in terms of linear transformations (instead of matrices).

(d) Generalize both (b) and (c) to larger numbers of pieces (instead of four).

§ 39. Invariance

A possible relation between subspaces $\mathfrak{M}$ of a vector space and linear transformations A on that space is invariance. We say that $\mathfrak{M}$ is *invariant* under A if x in $\mathfrak{M}$ implies that Ax is in $\mathfrak{M}$. (Observe that the implication relation is required in one direction only; we do not assume that every y in $\mathfrak{M}$ can be written in the form $y = Ax$ with x in $\mathfrak{M}$; we do not even assume that Ax in $\mathfrak{M}$ implies x in $\mathfrak{M}$. Presently we shall see examples in which the conditions we did not assume definitely fail to hold.) We know that a subspace of a vector space is itself a vector space; if we know that $\mathfrak{M}$ is invariant under A, we may ignore the fact that A is defined outside $\mathfrak{M}$ and we may consider A as a linear transformation defined on the vector space $\mathfrak{M}$. Invariance is often considered for sets of linear transformations, as well as for a single one; $\mathfrak{M}$ is invariant under a set if it is invariant under each member of the set.

What can be said about the matrix of a linear transformation A on an n-dimensional vector space $\mathfrak{V}$ if we know that some $\mathfrak{M}$ is invariant under A? In other words: is there a clever way of selecting a basis $\mathfrak{X} = \{x_1, \cdots, x_n\}$ in $\mathfrak{V}$ so that $[A] = [A; \mathfrak{X}]$ will have some particularly simple form? The answer is in § 12, Theorem 2; we may choose $\mathfrak{X}$ so that $x_1, \cdots, x_m$ are in $\mathfrak{M}$ and $x_{m+1}, \cdots, x_n$ are not. Let us express Ax_j in terms of $x_1, \cdots, x_n$. For $m + 1 \leq j \leq n$, there is not much we can say: $Ax_j = \sum_i \alpha_{ij} x_i$. For

$1 \leq j \leq m$, however, x_j is in $\mathfrak{M}$, and therefore (since $\mathfrak{M}$ is invariant under A)Ax_j is in $\mathfrak{M}$. Consequently, in this case Ax_j is a linear combination of $x_1, \cdots, x_m$; the α_{ij} with $m + 1 \leq i \leq n$ are zero. Hence the matrix $[A]$ of A, in this coordinate system, will have the form

$$[A] = \begin{pmatrix} [A_1] & [B_0] \\ [0] & [A_2] \end{pmatrix},$$

where $[A_1]$ is the (m-rowed) matrix of A considered as a linear transformation on the space $\mathfrak{M}$ (with respect to the coordinate system $\{x_1, \cdots, x_m\}$), $[A_2]$ and $[B_0]$ are some arrays of scalars (in size $(n - m)$ by $(n - m)$ and m by $(n - m)$ respectively), and $[0]$ denotes the rectangular $((n - m)$ by $m)$ array consisting of zeros only. (It is important to observe the unpleasant fact that $[B_0]$ need not be zero.)

§ 40. Reducibility

A particularly important subcase of the notion of invariance is that of reducibility. If $\mathfrak{M}$ and $\mathfrak{N}$ are two subspaces such that both are invariant under A and such that $\mathfrak{V}$ is their direct sum, then A is *reduced* (decomposed) by the pair $(\mathfrak{M}, \mathfrak{N})$. The difference between invariance and reducibility is that, in the former case, among the collection of all subspaces invariant under A we may not be able to pick out any two, other than $\mathfrak{O}$ and $\mathfrak{V}$, with the property that $\mathfrak{V}$ is their direct sum. Or, saying it the other way, if $\mathfrak{M}$ is invariant under A, there are, to be sure, many ways of finding an $\mathfrak{N}$ such that $\mathfrak{V} = \mathfrak{M} \oplus \mathfrak{N}$, but it may happen that no such $\mathfrak{N}$ will be invariant under A.

The process described above may also be turned around. Let $\mathfrak{M}$ and $\mathfrak{N}$ be any two vector spaces, and let A and B be any two linear transformations (on $\mathfrak{M}$ and $\mathfrak{N}$ respectively). Let $\mathfrak{V}$ be the direct sum $\mathfrak{M} \oplus \mathfrak{N}$; we may define on $\mathfrak{V}$ a linear transformation C called the direct sum of A and B, by writing

$$Cz = C(x, y) = (Ax, By).$$

We shall omit the detailed discussion of direct sums of transformations; we shall merely mention the results. Their proofs are easy. If $(\mathfrak{M}, \mathfrak{N})$ reduces C, and if we denote by A the linear transformation C considered on $\mathfrak{M}$ alone, and by B the linear transformation C considered on $\mathfrak{N}$ alone, then C is the direct sum of A and B. By suitable choice of basis (namely, by choosing $x_1, \cdots, x_m$ in $\mathfrak{M}$ and $x_{m+1}, \cdots, x_n$ in $\mathfrak{N}$) we may put the matrix of the direct sum of A and B in the form displayed in the preceding section, with $[A_1] = [A]$, $[B_0] = [0]$, and $[A_2] = [B]$. If p is any polynomial, and if we write $A' = p(A)$, $B' = p(B)$, then the direct sum C' of A' and B' will be $p(C)$.

1. Suppose that the matrix of a linear transformation (on a two-dimensiona vector space) with respect to some coordinate system is $\begin{pmatrix} 0 & 0 \\ 0 & 1 \end{pmatrix}$. How many subspaces are there invariant under the transformation?

2. Give an example of a linear transformation A on a finite-dimensional vector space $\mathcal{U}$ such that $\mathcal{O}$ and $\mathcal{U}$ are the only subspaces invariant under A.

3. Let D be the differentiation operator on $\mathcal{P}_n$. If $m \leq n$, then the subspace $\mathcal{P}_m$ is invariant under D. Is D on $\mathcal{P}_m$ invertible? Is there a complement of $\mathcal{P}_m$ in $\mathcal{P}_n$ such that it together with $\mathcal{P}_m$ reduces D?

4. Prove that the subspace spanned by two subspaces, each of which is invariant under some linear transformation A, is itself invariant under A.

§ 41. Projections

Especially important for our purposes is another connection between direct sums and linear transformations.

DEFINITION. If $\mathcal{U}$ is the direct sum of $\mathfrak{M}$ and $\mathfrak{N}$, so that every z in $\mathcal{U}$ may be written, uniquely, in the form $z = x + y$, with x in $\mathfrak{M}$ and y in $\mathfrak{N}$, the *projection* on $\mathfrak{M}$ along $\mathfrak{N}$ is the transformation E defined by $Ez = x$.

If direct sums are important, then projections are also, since, as we shall see, they are a very powerful algebraic tool in studying the geometric concept of direct sum. The reader will easily satisfy himself about the reason for the word "projection" by drawing a pair of axes (linear manifolds) in the plane (their direct sum). To make the picture look general enough, do not draw perpendicular axes!

We skipped over one point whose proof is easy enough to skip over, but whose existence should be recognized; it must be shown that E is a *linear transformation*. We leave this verification to the reader, and go on to look for special properties of projections.

THEOREM 1. *A linear transformation E is a projection on some subspace if and only if it is idempotent, that is, $E^2 = E$.*

PROOF. If E is the projection on $\mathfrak{M}$ along $\mathfrak{N}$, and if $z = x + y$, with x in $\mathfrak{M}$ and y in $\mathfrak{N}$, then the decomposition of x is $x + 0$, so that

$$E^2 z = EEz = Ex = x = Ez.$$

Conversely, suppose that $E^2 = E$. Let $\mathfrak{N}$ be the set of all vectors z in $\mathcal{U}$ for which $Ez = 0$; let $\mathfrak{M}$ be the set of all vectors z for which $Ez = z$. It is clear that both $\mathfrak{M}$ and $\mathfrak{N}$ are subspaces; we shall prove that $\mathcal{U} = \mathfrak{M}$

$\oplus$ $\mathfrak{N}$. In view of the theorem of § 18, we need to prove that $\mathfrak{M}$ and $\mathfrak{N}$ are disjoint and that together they span $\mathfrak{V}$.

If z is in $\mathfrak{M}$, then $Ez = z$; if z is in $\mathfrak{N}$, then $Ez = 0$; hence if z is in both $\mathfrak{M}$ and $\mathfrak{N}$, then $z = 0$. For an arbitrary z we have

$$z = Ez + (1 - E)z.$$

If we write $Ez = x$ and $(1 - E)z = y$, then $Ex = E^2z = Ez = x$, and $Ey = E(1 - E)z = Ez - E^2z = 0$, so that x is in $\mathfrak{M}$ and y is in $\mathfrak{N}$. This proves that $\mathfrak{V} = \mathfrak{M} \oplus \mathfrak{N}$, and that the projection on $\mathfrak{M}$ along $\mathfrak{N}$ is precisely E.

As an immediate consequence of the above proof we obtain also the following result.

Tʜᴇᴏʀᴇᴍ 2. *If E is the projection on $\mathfrak{M}$ along $\mathfrak{N}$, then $\mathfrak{M}$ and $\mathfrak{N}$ are, respectively, the sets of all solutions of the equations $Ez = z$ and $Ez = 0$.*

By means of these two theorems we can remove the apparent asymmetry, in the definition of projections, between the roles played by $\mathfrak{M}$ and $\mathfrak{N}$. If to every $z = x + y$ we make correspond not x but y, we also get an idempotent linear transformation. This transformation (namely, $1 - E$) is the projection on $\mathfrak{N}$ along $\mathfrak{M}$. We sum up the facts as follows.

Tʜᴇᴏʀᴇᴍ 3. *A linear transformation E is a projection if and only if $1 - E$ is a projection; if E is the projection on $\mathfrak{M}$ along $\mathfrak{N}$, then $1 - E$ is the projection on $\mathfrak{N}$ along $\mathfrak{M}$.*

§ 42. Combinations of projections

Continuing in the spirit of Theorem 3 of the preceding section, we investigate conditions under which various algebraic combinations of projections are themselves projections.

Tʜᴇᴏʀᴇᴍ. *We assume that E_1 and E_2 are projections on $\mathfrak{M}_1$ and $\mathfrak{M}_2$ along $\mathfrak{N}_1$ and $\mathfrak{N}_2$ respectively and that the underlying field of scalars is such that $1 + 1 \neq 0$. We make three assertions.*

(i) *$E_1 + E_2$ is a projection if and only if $E_1E_2 = E_2E_1 = 0$; if this condition is satisfied, then $E = E_1 + E_2$ is the projection on $\mathfrak{M}$ along $\mathfrak{N}$, where $\mathfrak{M} = \mathfrak{M}_1 \oplus \mathfrak{M}_2$ and $\mathfrak{N} = \mathfrak{N}_1 \cap \mathfrak{N}_2$.*

(ii) *$E_1 - E_2$ is a projection if and only if $E_1E_2 = E_2E_1 = E_2$; if this condition is satisfied, then $E = E_1 - E_2$ is the projection on $\mathfrak{M}$ along $\mathfrak{N}$, where $\mathfrak{M} = \mathfrak{M}_1 \cap \mathfrak{N}_2$ and $\mathfrak{N} = \mathfrak{N}_1 \oplus \mathfrak{M}_2$.*

(iii) *If $E_1E_2 = E_2E_1 = E$, then E is the projection on $\mathfrak{M}$ along $\mathfrak{N}$, where $\mathfrak{M} = \mathfrak{M}_1 \cap \mathfrak{M}_2$ and $\mathfrak{N} = \mathfrak{N}_1 + \mathfrak{N}_2$.*

PROOF. We recall the notation. If $\mathfrak{IC}$ and $\mathfrak{K}$ are subspaces, then $\mathfrak{IC} + \mathfrak{K}$ is the subspace spanned by $\mathfrak{IC}$ and $\mathfrak{K}$; writing $\mathfrak{IC} \oplus \mathfrak{K}$ implies that $\mathfrak{IC}$ and $\mathfrak{K}$ are disjoint, and then $\mathfrak{IC} \oplus \mathfrak{K} = \mathfrak{IC} + \mathfrak{K}$; and $\mathfrak{IC} \cap \mathfrak{K}$ is the intersection of $\mathfrak{IC}$ and $\mathfrak{K}$.

(i) If $E_1 + E_2 = E$ is a projection, then $(E_1 + E_2)^2 = E^2 = E = E_1 + E_2$, so that the cross-product terms must disappear:

$$(1) \qquad\qquad E_1E_2 + E_2E_1 = 0.$$

If we multiply (1) on both left and right by E_1, we obtain

$$E_1E_2 + E_1E_2E_1 = 0,$$

$$E_1E_2E_1 + E_2E_1 = 0;$$

subtracting, we get $E_1E_2 - E_2E_1 = 0$. Hence E_1 and E_2 are commutative, and (1) implies that their product is zero. (Here is where we need the assumption $1 + 1 \neq 0$.) Since, conversely, $E_1E_2 = E_2E_1 = 0$ clearly implies (1), we see that the condition is also sufficient to ensure that E be a projection.

Let us suppose, from now on, that E is a projection; by § 41, Theorem 2, $\mathfrak{M}$ and $\mathfrak{N}$ are, respectively, the sets of all solutions of the equations $Ez = z$ and $Ez = 0$. Let us write $z = x_1 + y_1 = x_2 + y_2$, where $x_1 = E_1z$ and $x_2 = E_2z$ are in $\mathfrak{M}_1$ and $\mathfrak{M}_2$, respectively, and $y_1 = (1 - E_1)z$ and $y_2 = (1 - E_2)z$ are in $\mathfrak{N}_1$ and $\mathfrak{N}_2$, respectively. If z is in $\mathfrak{M}$, $E_1z + E_2z = z$, then

$$z = E_1(x_2 + y_2) + E_2(x_1 + y_1) = E_1y_2 + E_2y_1.$$

Since $E_1(E_1y_2) = E_1y_2$ and $E_2(E_2y_1) = E_2y_1$, we have exhibited z as a sum of a vector from $\mathfrak{M}_1$ and a vector from $\mathfrak{M}_2$, so that $\mathfrak{M} \subset \mathfrak{M}_1 + \mathfrak{M}_2$. Conversely, if z is a sum of a vector from $\mathfrak{M}_1$ and a vector from $\mathfrak{M}_2$, then $(E_1 + E_2)z = z$, so that z is in $\mathfrak{M}$, and consequently $\mathfrak{M} = \mathfrak{M}_1 + \mathfrak{M}_2$. Finally, if z belongs to both $\mathfrak{M}_1$ and $\mathfrak{M}_2$, so that $E_1z = E_2z = z$, then $z = E_1z = E_1(E_2z) = 0$, so that $\mathfrak{M}_1$ and $\mathfrak{M}_2$ are disjoint; we have proved that $\mathfrak{M} = \mathfrak{M}_1 \oplus \mathfrak{M}_2$.

It remains to find $\mathfrak{N}$, that is, to find all solutions of $E_1z + E_2z = 0$. If z is in $\mathfrak{N}_1 \cap \mathfrak{N}_2$, this equation is clearly satisfied; conversely $E_1z + E_2z = 0$ implies (upon multiplication on the left by E_1 and E_2 respectively) that $E_1z + E_1E_2z = 0$ and $E_2E_1z + E_2z = 0$. Since $E_1E_2z = E_2E_1z = 0$ for all z, we obtain finally $E_1z = E_2z = 0$, so that z belongs to both $\mathfrak{N}_1$ and $\mathfrak{N}_2$.

With the technique and the results obtained in this proof, the proofs of the remaining parts of the theorem are easy.

(ii) According to § 41, Theorem 3, $E_1 - E_2$ is a projection if and only if $1 - (E_1 - E_2) = (1 - E_1) + E_2$ is a projection. According to (i) this happens (since, of course, $1 - E_1$ is the projection on $\mathfrak{N}_1$ along $\mathfrak{M}_1$) if and only if

(2) $$(1 - E_1)E_2 = E_2(1 - E_1) = 0,$$

and in this case $(1 - E_1) + E_2$ is the projection on $\mathfrak{N}_1 \oplus \mathfrak{M}_2$ along $\mathfrak{M}_1 \cap \mathfrak{N}_2$. Since (2) is equivalent to $E_1E_2 = E_2E_1 = E_2$, the proof of (ii) is complete.

(iii) That $E = E_1E_2 = E_2E_1$ implies that E is a projection is clear, since E is idempotent. We assume, therefore, that E_1 and E_2 commute and we find $\mathfrak{M}$ and $\mathfrak{N}$. If $Ez = z$, then $E_1z = E_1Ez = E_1E_1E_2z = E_1E_2z = z$, and similarly $E_2z = z$, so that z is contained in both $\mathfrak{M}_1$ and $\mathfrak{M}_2$. The converse is clear; if $E_1z = z = E_2z$, then $Ez = z$. Suppose next that $E_1E_2z = 0$; it follows that E_2z belongs to $\mathfrak{N}_1$, and, from the commutativity of E_1 and E_2, that E_1z belongs to $\mathfrak{N}_2$. This is more symmetry than we need; since $z = E_2z + (1 - E_2)z$, and since $(1 - E_2)z$ is in $\mathfrak{N}_2$, we have exhibited z as a sum of a vector from $\mathfrak{N}_1$ and a vector from $\mathfrak{N}_2$. Conversely if z is such a sum, then $E_1E_2z = 0$; this concludes the proof that $\mathfrak{N} = \mathfrak{N}_1 + \mathfrak{N}_2$.

We shall return to theorems of this type later, and we shall obtain, in certain cases, more precise results. Before leaving the subject, however, we call attention to a few minor peculiarities of the theorem of this section. We observe first that although in both (i) and (ii) one of $\mathfrak{M}$ and $\mathfrak{N}$ was a *direct* sum of the given subspaces, in (iii) we stated only that $\mathfrak{N} = \mathfrak{N}_1 + \mathfrak{N}_2$. Consideration of the possibility $E_1 = E_2 = E$ shows that this is unavoidable. Also: the condition of (iii) was asserted to be sufficient only; it is possible to construct projections E_1 and E_2 whose product E_1E_2 is a projection, but for which E_1E_2 and E_2E_1 are different. Finally, it may be conjectured that it is possible to extend the result of (i), by induction, to more than two summands. Although this is true, it is surprisingly non-trivial; we shall prove it later in a special case of interest.

§ 43. Projections and invariance

We have already seen that the study of projections is equivalent to the study of direct sum decompositions. By means of projections we may also study the notions of invariance and reducibility.

THEOREM 1. *If a subspace $\mathfrak{M}$ is invariant under the linear transformation A, then $EAE = AE$ for every projection E on $\mathfrak{M}$. Conversely, if $EAE = AE$ for some projection E on $\mathfrak{M}$, then $\mathfrak{M}$ is invariant under A.*

PROOF. Suppose that $\mathfrak{M}$ is invariant under A and that $\mathcal{U} = \mathfrak{M} \oplus \mathfrak{N}$ for some $\mathfrak{N}$; let E be the projection on $\mathfrak{M}$ along $\mathfrak{N}$. For any $z = x + y$

(with x in $\mathfrak{M}$ and y in $\mathfrak{N}$) we have $AEz = Ax$ and $EAEz = EAx$; since the presence of x in $\mathfrak{M}$ guarantees the presence of Ax in $\mathfrak{M}$, it follows that EAx is also equal to Ax, as desired.

Conversely, suppose that $\mathfrak{V} = \mathfrak{M} \oplus \mathfrak{N}$, and that $EAE = AE$ for the projection E on $\mathfrak{M}$ along $\mathfrak{N}$. If x is in $\mathfrak{M}$, then $Ex = x$, so that

$$EAx = EAEx = AEx = Ax,$$

and consequently Ax is also in $\mathfrak{M}$.

THEOREM 2. *If $\mathfrak{M}$ and $\mathfrak{N}$ are subspaces with $\mathfrak{V} = \mathfrak{M} \oplus \mathfrak{N}$, then a necessary and sufficient condition that the linear transformation A be reduced by the pair $(\mathfrak{M}, \mathfrak{N})$ is that $EA = AE$, where E is the projection on $\mathfrak{M}$ along $\mathfrak{N}$.*

PROOF. First we assume that $EA = AE$, and we prove that A is reduced by $(\mathfrak{M}, \mathfrak{N})$. If x is in $\mathfrak{M}$, then $Ax = AEx = EAx$, so that Ax is also in $\mathfrak{M}$; if x is in $\mathfrak{N}$, then $Ex = 0$ and $EAx = AEx = A0 = 0$, so that Ax is also in $\mathfrak{N}$.

Next we assume that A is reduced by $(\mathfrak{M}, \mathfrak{N})$, and we prove that $EA = AE$. Since $\mathfrak{M}$ is invariant under A, Theorem 1 assures us that $EAE = AE$; since $\mathfrak{N}$ is also invariant under A, and since $1 - E$ is a projection on $\mathfrak{N}$, we have, similarly, $(1 - E)A(1 - E) = A(1 - E)$. From the second equation, after carrying out the indicated multiplications and simplifying, we obtain $EAE = EA$; this concludes the proof of the theorem.

EXERCISES

1. (a) Suppose that E is a projection on a vector space $\mathfrak{V}$, and suppose that scalar multiplication is redefined so that the new product of a scalar α and a vector x is the old product of α and Ex. Show that vector addition (old) and scalar multiplication (new) satisfy all the axioms on a vector space except $1 \cdot x = x$.

(b) To what extent is it true that the method described in (a) is the only way to construct systems satisfying all the axioms on a vector space except $1 \cdot x = x$?

2. (a) Suppose that $\mathfrak{V}$ is a vector space, x_0 is a vector in $\mathfrak{V}$, and y_0 is a linear functional on $\mathfrak{V}$; write $Ax = [x, y_0]x_0$ for every x in $\mathfrak{V}$. Under what conditions on x_0 and y_0 is A a projection?

(b) If A is the projection on, say, $\mathfrak{M}$ along $\mathfrak{N}$, characterize $\mathfrak{M}$ and $\mathfrak{N}$ in terms of x_0 and y_0.

3. If A is left multiplication by P on a space of linear transformations (cf. § 38 Ex. 5), under what conditions on P is A a projection?

4. If A is a linear transformation, if E is a projection, and if $F = 1 - E$, then

$$A = EAE + EAF + FAE + FAF.$$

Use this result to prove the multiplication rule for partitioned (square) matrices (as in § 38, Ex. 19).

5. (a) If E_1 and E_2 are projections on $\mathfrak{M}_1$ and $\mathfrak{M}_2$ along $\mathfrak{N}_1$ and $\mathfrak{N}_2$ respectively, and if E_1 and E_2 commute, then $E_1 + E_2 - E_1E_2$ is a projection.

(b) If $E_1 + E_2 - E_1E_2$ is the projection on $\mathfrak{M}$ along $\mathfrak{N}$, describe $\mathfrak{M}$ and $\mathfrak{N}$ in terms of $\mathfrak{M}_1$, $\mathfrak{M}_2$, $\mathfrak{N}_1$, and $\mathfrak{N}_2$.

6. (a) Find a linear transformation A such that $A^2(1 - A) = 0$ but A is not idempotent.

(b) Find a linear transformation A such that $A(1 - A)^2 = 0$ but A is not idempotent.

(c) Prove that if A is a linear transformation such that $A^2(1 - A) = A(1 - A)^2 = 0$, then A is idempotent.

7. (a) Prove that if E is a projection on a finite-dimensional vector space, then there exists a basis $\mathfrak{X}$ such that the matrix (e_{ij}) of E with respect to $\mathfrak{X}$ has the following special form: $e_{ij} = 0$ or 1 for all i and j, and $e_{ij} = 0$ if $i \neq j$.

(b) An *involution* is a linear transformation U such that $U^2 = 1$. Show that if $1 + 1 \neq 0$, then the equation $U = 2E - 1$ establishes a one-to-one correspondence between all projections E and all involutions U.

(c) What do (a) and (b) imply about the matrix of an involution on a finite-dimensional vector space?

8. (a) In the space $\mathbb{C}^2$ of all vectors (ξ_1, ξ_2) let $\mathfrak{M}^+$, $\mathfrak{N}_1$, and $\mathfrak{N}_2$ be the subspaces characterized by $\xi_1 = \xi_2$, $\xi_1 = 0$, and $\xi_2 = 0$, respectively. If E_1 and E_2 are the projections on $\mathfrak{M}^+$ along $\mathfrak{N}_1$ and $\mathfrak{N}_2$ respectively, show that $E_1E_2 = E_2$ and $E_2E_1 = E_1$.

(b) Let $\mathfrak{M}^-$ be the subspace characterized by $\xi_1 = -\xi_2$. If E_0 is the projection on $\mathfrak{N}_2$ along $\mathfrak{M}^-$, then E_2E_0 is a projection, but E_0E_2 is not.

9. Show that if E, F, and G are projections on a vector space over a field whose characteristic is not equal to 2, and if $E + F + G = 1$, then $EF = FE = EG = GE = FG = GF = 0$. Does the proof work for four projections instead of three?

§ 44. Adjoints

Let us study next the relation between the notions of linear transformation and dual space. Let $\mathcal{U}$ be any vector space and let y be any element of $\mathcal{U}'$; for any linear transformation A on $\mathcal{U}$ we consider the expression $[Ax, y]$. For each fixed y, the function y' defined by $y'(x) = [Ax, y]$ is a linear functional on $\mathcal{U}$; using the square bracket notation for y' as well as for y, we have $[Ax, y] = [x, y']$. If now we allow y to vary over $\mathcal{U}'$, then this procedure makes correspond to each y a y', depending, of course, on y; we write $y' = A'y$. The defining property of A' is

$$(1) \qquad\qquad [Ax, y] = [x, A'y].$$

We assert that A' is a linear transformation on $\mathcal{U}'$. Indeed, if $y = \alpha_1 y_1 + \alpha_2 y_2$, then

$$[x, A'y] = [Ax, y] = \alpha_1[Ax, y_1] + \alpha_2[Ax, y_2]$$
$$= \alpha_1[x, A'y_1] + \alpha_2[x, A'y_2] = [x, \alpha_1 A'y_1 + \alpha_2 A'y_2].$$

The linear transformation A' is called the *adjoint* (or dual) of A; we dedicate this section and the next to studying properties of A'. Let us first get the formal algebraic rules out of the way; they go as follows.

(2) $$0' = 0,$$

(3) $$1' = 1,$$

(4) $$(A + B)' = A' + B',$$

(5) $$(\alpha A)' = \alpha A',$$

(6) $$(AB)' = B'A',$$

(7) $$(A^{-1})' = (A')^{-1}.$$

Here (7) is to be interpreted in the following sense: if A is invertible, then so is A', and the equation is valid. The proofs of all these relations are elementary; to indicate the procedure, we carry out the computations for (6) and (7). To prove (6), merely observe that

$$[ABx, y] = [Bx, A'y] = [x, B'A'y].$$

To prove (7), suppose that A is invertible, so that $AA^{-1} = A^{-1}A = 1$. Applying (3) and (6) to these equations, we obtain

$$(A^{-1})'A' = A'(A^{-1})' = 1;$$

Theorem 1 of § 36 implies that A' is invertible and that (7) is valid.

In finite-dimensional spaces another important relation holds:

(8) $$A'' = A.$$

This relation has to be read with a grain of salt. As it stands A'' is a transformation not on $\mathcal{U}$ but on the dual space $\mathcal{U}''$ of $\mathcal{U}'$. If, however, we identify $\mathcal{U}''$ and $\mathcal{U}$ according to the natural isomorphism, then A'' acts on $\mathcal{U}$ and (8) makes sense. In this interpretation the proof of (8) is trivial. Since $\mathcal{U}$ is reflexive, we obtain every linear functional on $\mathcal{U}'$ by considering $[x, y]$ as a function of y, with x fixed in $\mathcal{U}$. Since $[x, A'y]$ defines a function (a linear functional) of y, it may be written in the form $[x', y]$. The vector x' here is, by definition, $A''x$. Hence we have, for every y in $\mathcal{U}'$ and for every x in $\mathcal{U}$,

$$[Ax, y] = [x, A'y] = [A''x, y];$$

the equality of the first and last terms of this chain proves (8).

Under the hypothesis of (8) (that is, finite-dimensionality), the asymmetry in the interpretation of (7) may be removed; we assert that in this case the invertibility of A' implies that of A and, therefore, the validity of (7). Proof: apply the old interpretation of (7) to A' and A'' in place of A and A'.

Our discussion is summed up, in the reflexive finite-dimensional case, by the assertion that the mapping $A \rightarrow A'$ is one-to-one, and, in fact, an algebraic anti-isomorphism, from the set of all linear transformations on $\mathcal{U}$ onto the set of all linear transformations on $\mathcal{U}'$. (The prefix "anti" got attached because of the commutation rule (6).)

§ 45. Adjoints of projections

There is one important case in which multiplication does not get turned around, that is, when $(AB)' = A'B'$; namely, the case when A and B commute. We have, in particular, $(A^n)' = (A')^n$, and, more generally, $(p(A))' = p(A')$ for every polynomial p. It follows from this that if E is a projection, then so is E'. The question arises: what direct sum decomposition is E' associated with?

THEOREM 1. *If E is the projection on $\mathfrak{M}$ along $\mathfrak{N}$, then E' is the projection on $\mathfrak{N}^0$ along $\mathfrak{M}^0$.*

PROOF. We know already that $(E')^2 = E'$ and $\mathcal{U}' = \mathfrak{N}^0 \oplus \mathfrak{M}^0$ (cf. § 20). It is necessary only to find the subspaces consisting of the solutions of $E'y = 0$ and $E'y = y$. This we do in four steps.

(i) If y is in $\mathfrak{M}^0$, then, for all x,

$$[x, E'y] = [Ex, y] = 0,$$

so that $E'y = 0$.

(ii) If $E'y = 0$, then, for all x in $\mathfrak{M}$,

$$[x, y] = [Ex, y] = [x, E'y] = 0,$$

so that y is in $\mathfrak{M}^0$.

(iii) If y is in $\mathfrak{N}^0$, then, for all x,

$$[x, y] = [Ex, y] + [(1 - E)x, y] = [Ex, y] = [x, E'y],$$

so that $E'y = y$.

(iv) If $E'y = y$, then for all x in $\mathfrak{N}$,

$$[x, y] = [x, E'y] = [Ex, y] = 0,$$

so that y is in $\mathfrak{N}^0$.

Steps (i) and (ii) together show that the set of solutions of $E'y = 0$ is precisely $\mathfrak{M}^0$; steps (iii) and (iv) together show that the set of solutions of $E'y = y$ is precisely $\mathfrak{N}^0$. This concludes the proof of the theorem.

THEOREM 2. *If $\mathfrak{M}$ is invariant under A, then $\mathfrak{M}^0$ is invariant under A'; if A is reduced by $(\mathfrak{M}, \mathfrak{N})$, then A' is reduced by $(\mathfrak{M}^0, \mathfrak{N}^0)$.*

PROOF. We shall prove only the first statement; the second one clearly follows from it. We first observe the following identity, valid for any three linear transformations E, F, and A, subject to the relation $F = 1 - E$:

$$(1) \qquad FAF - FA = EAE - AE.$$

(Compare this with the proof of § 43, Theorem 2.) Let E be any projection on $\mathfrak{M}$; by § 43, Theorem 1, the right member of (1) vanishes, and, therefore, so does the left member. By taking adjoints, we obtain $F'A'F' = A'F'$; since, by Theorem 1 of the present section, $F' = 1 - E'$ is a projection on $\mathfrak{M}^0$, the proof of Theorem 2 is complete. (Here is an alternative proof of the first statement of Theorem 2, a proof that does not make use of the fact that $\mathbb{U}$ is the direct sum of $\mathfrak{M}$ and some other subspace. If y is in $\mathfrak{M}^0$, then $[x, A'y] = [Ax, y] = 0$ for all x in $\mathfrak{M}$, and therefore $A'y$ is in $\mathfrak{M}^0$. The only advantage of the algebraic proof given above over this simple geometric proof is that the former prepares the ground for future work with projections.)

We conclude our treatment of adjoints by discussing their matrices; this discussion is intended to illuminate the entire theory and to enable the reader to construct many examples.

We shall need the following fact: if $\mathfrak{X} = \{x_1, \cdots, x_n\}$ is any basis in the n-dimensional vector space $\mathbb{U}$, if $\mathfrak{X}' = \{y_1, \cdots, y_n\}$ is the dual basis in $\mathbb{U}'$, and if the matrix of the linear transformation A in the coordinate system $\mathfrak{X}$ is (α_{ij}), then

$$(2) \qquad \alpha_{ij} = [Ax_j, y_i].$$

This follows from the definition of the matrix of a linear transformation; since $Ax_j = \sum_k \alpha_{kj} x_k$, we have

$$[Ax_j, y_i] = \sum_k \alpha_{kj}[x_k, y_i] = \alpha_{ij}.$$

To keep things straight in the applications, we rephrase formula (2) verbally, thus: to find the (i, j) element of $[A]$ in the basis $\mathfrak{X}$, apply A to the j-th element of $\mathfrak{X}$ and then take the value of the i-th linear functional (in $\mathfrak{X}'$) at the vector so obtained.

It is now very easy to find the matrix $(\alpha'_{ij}) = [A']$ in the coordinate system $\mathfrak{X}'$; we merely follow the recipe just given. In other words, we consider $A'y_j$, and take the value of the i-th linear functional in $\mathfrak{X}''$ (that is, of x_i considered as a linear functional on $\mathfrak{X}'$) at this vector; the result is that

$$\alpha'_{ij} = [x_i, A'y_j].$$

Since $[x_i, A'y_j] = [Ax_i, y_j] = \alpha_{ji}$, so that $\alpha'_{ij} = \alpha_{ji}$, this matrix $[A']$ is called the *transpose* of $[A]$.

Observe that our results on the relation between E and E' (where E is a projection) could also have been derived by using the facts about the matricial representation of a projection together with the present result on the matrices of adjoint transformations.

§ 46. Change of basis

Although what we have been doing with linear transformations so far may have been complicated, it was to a large extent automatic. Having introduced the new concept of linear transformation, we merely let some of the preceding concepts suggest ways in which they are connected with linear transformations. We now begin the proper study of linear transformations. As a first application of the theory we shall solve the problems arising from a change of basis. These problems can be formulated without mentioning linear transformations, but their solution is most effectively given in terms of linear transformations.

Let $\mathcal{V}$ be an n-dimensional vector space and let $\mathcal{X} = \{x_1, \cdots, x_n\}$ and $\mathcal{Y} = \{y_1, \cdots, y_n\}$ be two bases in $\mathcal{V}$. We may ask the following two questions.

QUESTION I. *If x is in $\mathcal{V}$, $x = \sum_i \xi_i x_i = \sum_i \eta_i y_i$, what is the relation between its coordinates $(\xi_1, \cdots, \xi_n)$ with respect to $\mathcal{X}$ and its coordinates $(\eta_1, \cdots, \eta_n)$ with respect to $\mathcal{Y}$?*

QUESTION II. *If $(\xi_1, \cdots, \xi_n)$ is an ordered set of n scalars, what is the relation between the vectors $x = \sum_i \xi_i x_i$ and $y = \sum_i \xi_i y_i$?*

Both these questions are easily answered in the language of linear transformations. We consider, namely, the linear transformation A defined by $Ax_i = y_i, i = 1, \cdots, n$. More explicitly:

$$A\left(\sum_i \xi_i x_i\right) = \sum_i \xi_i y_i.$$

Let (α_{ij}) be the matrix of A in the basis $\mathcal{X}$, that is, $y_j = Ax_j = \sum_i \alpha_{ij} x_i$. We observe that A is invertible, since $\sum_i \xi_i y_i = 0$ implies that $\xi_1 = \xi_2 = \cdots = \xi_n = 0$.

ANSWER TO QUESTION I. Since

$$\sum_j \eta_j y_j = \sum_j \eta_j A x_j = \sum_j \eta_j \sum_i \alpha_{ij} x_i$$
$$= \sum_i \left(\sum_j \alpha_{ij} \eta_j\right) x_i,$$

we have

$$(1) \qquad \qquad \xi_i = \sum_j \alpha_{ij} \eta_j.$$

ANSWER TO QUESTION II.

$$(2) \qquad \qquad y = Ax.$$

Roughly speaking, the invertible linear transformation A (or, more properly, the matrix (α_{ij})) may be considered as a transformation of coordinates (as in (1)), or it may be considered (as we usually consider it, in (2)) as a transformation of vectors.

In classical treatises on vector spaces it is customary to treat vectors as numerical n-tuples, rather than as abstract entities; this necessitates the introduction of some cumbersome terminology. We give here a brief glossary of some of the more baffling terms and notations that arise in connection with dual spaces and adjoint transformations.

If $\mathcal{V}$ is an n-dimensional vector space, a vector x is given by its coordinates with respect to some preferred, absolute coordinate system; these coordinates form an ordered set of n scalars. It is customary to write this set of scalars in a column,

$$x = \begin{bmatrix} \xi_1 \\ \cdot \\ \cdot \\ \cdot \\ \xi_n \end{bmatrix}.$$

Elements of the dual space $\mathcal{V}'$ are written as rows, $x' = (\xi'_1, \cdots, \xi'_n)$. If we think of x as a (rectangular) n-by-one matrix, and of x' as a one-by-n matrix, then the matrix product $x'x$ is a one-by-one matrix, that is, a scalar. In our notation this scalar is $[x, x'] = \xi_1\xi'_1 + \cdots + \xi_n\xi'_n$. The trick of considering vectors as thin matrices works even when we consider the full-grown matrices of linear transformations. Thus the matrix product of (α_{ij}) with the column (ξ_j) is the column whose i-th element is $\eta_i = \sum_j \alpha_{ij}\xi_j$. Instead of worrying about dual bases and adjoint transformations, we may form similarly the product of the row (ξ'_j) with the matrix (α_{ij}) in the order $(\xi'_j)(\alpha_{ij})$; the result is the row that we earlier denoted by $y' = A'x'$. The expression $[Ax, x']$ is now abbreviated as $x' \cdot A \cdot x$; both dots denote ordinary matrix multiplication. The vectors x in $\mathcal{V}$ are called *covariant* and the vectors x' in $\mathcal{V}'$ are called *contravariant*. Since the notion of the product $x' \cdot x$ (that is, $[x, x']$) depends, from this point of view, on the coordinates of x and x', it becomes relevant to ask the following question: if we change basis in $\mathcal{V}$, in accordance with the invertible linear transformation A, what must we do in $\mathcal{V}'$ to preserve the product $x' \cdot x$? In our notation: if $[x, x'] = [y, y']$, where $y = Ax$, then how is y' related to x'? Answer: $y' = (A')^{-1}x'$. To express this whole tangle of ideas the classical terminology says that the vectors x vary *cogrediently* whereas the x' vary *contragrediently*.

§ 47. Similarity

The following two questions are closely related to those of the preceding section.

QUESTION III. *If B is a linear transformation on $\mathcal{V}$, what is the relation between its matrix (β_{ij}) with respect to $\mathfrak{X}$ and its matrix (γ_{ij}) with respect to $\mathcal{Y}$?*

QUESTION IV. *If (β_{ij}) is a matrix, what is the relation between the linear transformations B and C defined, respectively, by $Bx_j = \sum_i \beta_{ij}x_i$ and $Cy_j = \sum_i \beta_{ij}y_i$?*

Questions III and IV are explicit formulations of a problem we raised before: to one transformation there correspond (in different coordinate systems) many matrices (question III) and to one matrix there correspond many transformations (question IV).

ANSWER TO QUESTION III. We have

$$(1) \qquad\qquad Bx_j = \sum_i \beta_{ij}x_i$$

and

$$(2) \qquad\qquad By_j = \sum_i \gamma_{ij}y_i.$$

Using the linear transformation A defined in the preceding section, we may write

$$(3) \qquad By_j = BAx_j = B\left(\sum_k \alpha_{kj}x_k\right)$$
$$= \sum_k \alpha_{kj}Bx_k = \sum_k \alpha_{kj}\sum_i \beta_{ik}x_i = \sum_i \left(\sum_k \beta_{ik}\alpha_{kj}\right)x_i,$$

and

$$(4) \qquad \sum_k \gamma_{kj}y_k = \sum_k \gamma_{kj}Ax_k = \sum_k \gamma_{kj}\sum_i \alpha_{ik}x_i$$
$$= \sum_i \left(\sum_k \alpha_{ik}\gamma_{kj}\right)x_i.$$

Comparing (2), (3), and (4), we see that

$$\sum_k \alpha_{ik}\gamma_{kj} = \sum_k \beta_{ik}\alpha_{kj}.$$

Using matrix multiplication, we write this in the dangerously simple form

$$(5) \qquad\qquad [A][C] = [B][A].$$

The danger lies in the fact that three of the four matrices in (5) correspond to their linear transformations in the basis $\mathfrak{X}$; the fourth one—namely, the one we denoted by $[C]$—corresponds to B in the basis $\mathcal{Y}$. With this understanding, however, (5) is correct. A more usual form of (5), adapted, in principle, to computing $[C]$ when $[A]$ and $[B]$ are known, is

$$(6) \qquad\qquad [C] = [A]^{-1}[B][A].$$

ANSWER TO QUESTION IV. To bring out the essentially geometric character of this question and its answer, we observe that

$$Cy_j = CAx_j$$

and

$$\sum_i \beta_{ij} y_i = \sum_i \beta_{ij} A x_i = A(\sum_i \beta_{ij} x_i) = ABx_j.$$

Hence C is such that

$$CAx_j = ABx_j,$$

or, finally,

$$(7) \qquad\qquad C = ABA^{-1}.$$

There is no trouble with (7) similar to the one that caused us to make a reservation about the interpretation of (6); to find the linear transformation (not matrix) C, we multiply the transformations A, B, and A^{-1}, and nothing needs to be said about coordinate systems. Compare, however, the formulas (6) and (7), and observe once more the innate perversity of mathematical symbols. This is merely another aspect of the facts already noted in §§ 37 and 38.

Two matrices $[B]$ and $[C]$ are called *similar* if there exists an invertible matrix $[A]$ satisfying (6); two linear transformations B and C are called similar if there exists an invertible transformation A satisfying (7). In this language the answers to questions III and IV can be expressed very briefly; in both cases the answer is that the given matrices or transformations must be similar.

Having obtained the answer to question IV, we see now that there are too many subscripts in its formulation. The validity of (7) is a geometric fact quite independent of linearity, finite-dimensionality, or any other accidental property that A, B, and C may possess; the answer to question IV is also the answer to a much more general question. This geometric question, a paraphrase of the analytic formulation of question IV, is this: If B transforms v, and if C transforms Av the same way, what is the relation between B and C? The expression "the same way" is not so vague as it sounds; it means that if B takes x into, say, u, then C takes Ax into Au. The answer is, of course, the same as before: since $Bx = u$ and $Cy = v$ (where $y = Ax$ and $v = Au$), we have

$$ABx = Au = v = Cy = CAx.$$

The situation is conveniently summed up in the following mnemonic diagram:

$$\begin{array}{ccc} & B & \\ x & \longrightarrow & u \\ A \downarrow & & \downarrow A \\ y & \longrightarrow & v \\ & C & \end{array}$$

We may go from y to v by using the short cut C, or by going around the block; in other words $C = ABA^{-1}$. Remember that ABA^{-1} is to be applied to y from right to left: first A^{-1}, then B, then A.

We have seen that the theory of changing bases is coextensive with the theory of invertible linear transformations. An invertible linear transformation is an *automorphism*, where by an automorphism we mean an isomorphism of a vector space with itself. (See § 9.) We observe that, conversely, every automorphism is an invertible linear transformation.

We hope that the relation between linear transformations and matrices is by now sufficiently clear that the reader will not object if in the sequel, when we wish to give examples of linear transformations with various properties, we content ourselves with writing down a matrix. The interpretation always to be placed on this procedure is that we have in mind the concrete vector space $\mathbb{C}^n$ (or one of its generalized versions $\mathcal{F}^n$) and the concrete basis $\mathcal{X} = \{x_1, \cdots, x_n\}$ defined by $x_i = (\delta_{i1}, \cdots, \delta_{in})$. With this understanding, a matrix (α_{ij}) defines, of course, a unique linear transformation A, given by the usual formula $A(\sum_i \xi_i x_i) = \sum_i (\sum_j \alpha_{ij}\xi_j)x_i$.

EXERCISES

1. If A is a linear transformation from a vector space $\mathcal{U}$ to a vector space $\mathcal{V}$, then corresponding to each fixed y in $\mathcal{V}'$ there exists a vector, which might as well be denoted by $A'y$, in $\mathcal{U}'$ so that

$$[Ax, y] = [x, A'y]$$

for all x in $\mathcal{U}$. Prove that A' is a linear transformation from $\mathcal{V}'$ to $\mathcal{U}'$. (The transformation A' is called the *adjoint* of A.) Interpret and prove as many as possible among the equations § 44, (2)–(8) for this concept of adjoint.

2. (a) Prove that similarity of linear transformations on a vector space is an equivalence relation (that is, it is reflexive, symmetric, and transitive).

(b) If A is similar to a scalar α, then $A = \alpha$.

(c) If A and B are similar, then so also are A^2 and B^2, A' and B', and, in case A and B are invertible, A^{-1} and B^{-1}.

(d) Generalize the concept of similarity to two transformations defined on different vector spaces. Which of the preceding results remain valid for the generalized concept?

3. (a) If A and B are linear transformations on the same vector space and if at least one of them is invertible, then AB and BA are similar.

(b) Does the conclusion of (a) remain valid if neither A nor B is invertible?

4. If the matrix of a linear transformation A on $\mathbb{C}^2$, with respect to the basis $\{(1, 0), (0, 1)\}$ is $\begin{pmatrix} 1 & 1 \\ 1 & 1 \end{pmatrix}$, what is the matrix of A with respect to the basis $\{(1, 1)$ $(1, -1)\}$? What about the basis $\{(1, 0), (1, 1)\}$?

5. If the matrix of a linear transformation A on $\mathbb{C}^3$, with respect to the basis $\{(1, 0, 0), (0, 1, 0), (0, 0, 1)\}$ is $\begin{pmatrix} 0 & 1 & 1 \\ 1 & 0 & -1 \\ -1 & -1 & 0 \end{pmatrix}$, what is the matrix of A with respect to the basis $\{(0, 1, -1), (1, -1, 1), (-1, 1, 0)\}$?

6. (a) The construction of a matrix associated with a linear transformation depends on two bases, not one. Indeed, if $\mathfrak{X} = \{x_1, \cdots, x_n\}$ and $\bar{\mathfrak{X}} = \{\bar{x}_1, \cdots, \bar{x}_n\}$ are bases of $\mathcal{V}$, and if A is a linear transformation on V, then the matrix $[A; \mathfrak{X}, \bar{\mathfrak{X}}]$ of A with respect to $\mathfrak{X}$ and $\bar{\mathfrak{X}}$ should be defined by

$$Ax_j = \sum_i \alpha_{ij}\bar{x}_i.$$

The definition adopted in the text corresponds to the special case in which $\bar{\mathfrak{X}} = \mathfrak{X}$. The special case leads to the definition of similarity (B and C are similar if there exist bases $\mathfrak{X}$ and $\mathcal{Y}$ such that $[B; \mathfrak{X}] = [C; \mathcal{Y}]$). The analogous relation suggested by the general case is called equivalence; B and C are *equivalent* if there exist basis pairs $(\mathfrak{X}, \bar{\mathfrak{X}})$ and $(\mathcal{Y}, \bar{\mathcal{Y}})$ such that $[B; \mathfrak{X}, \bar{\mathfrak{X}}] = [C; \mathcal{Y}, \bar{\mathcal{Y}}]$. Prove that this notion is indeed an equivalence relation.

(b) Two linear transformations B and C are equivalent if and only if there exist invertible linear transformations P and Q such that $PB = CQ$.

(c) If A and B are equivalent, then so also are A' and B'.

(d) Does there exist a linear transformation A such that A is equivalent to a scalar α, but $A \neq \alpha$?

(e) Do there exist linear transformations A and B such that A and B are equivalent, but A^2 and B^2 are not?

(f) Generalize the concept of equivalence to two transformations defined on different vector spaces. Which of the preceding results remain valid for the generalized concept?

§ 48. Quotient transformations

Suppose that A is a linear transformation on a vector space $\mathcal{V}$ and that $\mathfrak{M}$ is a subspace of $\mathcal{V}$ invariant under A. Under these circumstances there is a natural way of defining a linear transformation (to be denoted by $A/\mathfrak{M}$) on the space $\mathcal{V}/\mathfrak{M}$; this "quotient transformation" is related to A just about the same way as the quotient space is related to $\mathcal{V}$. It will be convenient (in this section) to denote $\mathcal{V}/\mathfrak{M}$ by the more compact symbol $\mathcal{V}^-$, and to use related symbols for the vectors and the linear transformations that occur. Thus, for instance, if x is any vector in $\mathcal{V}$, we shall denote the coset $x + \mathfrak{M}$ by x^-; objects such as x^- are the typical elements of $\mathcal{V}^-$.

To define the quotient transformation $A/\mathfrak{M}$ (to be denoted, alternatively, by A^-), write

$$A^- x^- = (Ax)^-$$

for every vector x in $\mathcal{V}$. In other words, to find the transform by $A/\mathfrak{M}$ of the coset $x + \mathfrak{M}$, first find the transform by A of the vector x, and then form the coset of $\mathfrak{M}$ determined by that transformed vector. This definition must be supported by an unambiguity argument; we must be sure that if two vectors determine the same coset, then the same is true of their transforms by A. The key fact here is the invariance of $\mathfrak{M}$. Indeed, if $x + \mathfrak{M} = y + \mathfrak{M}$, then $x - y$ is in $\mathfrak{M}$, so that (invariance) $Ax - Ay$ is in $\mathfrak{M}$, and therefore $Ax + \mathfrak{M} = Ay + \mathfrak{M}$.

What happens if $\mathfrak{M}$ is not merely invariant under A, but, together with a suitable subspace $\mathfrak{N}$, reduces A? If this happens, then A is the direct sum, say $A = B \oplus C$, of two linear transformations defined on the subspaces $\mathfrak{M}$ and $\mathfrak{N}$ of $\mathcal{V}$, respectively; the question is, what is the relation between A^- and C? Both these transformations can be considered as complementary to A; the transformation B describes what A does on $\mathfrak{M}$, and both A^- and C describe in different ways what A does elsewhere.

Let T be the correspondence that assigns to each vector x in $\mathfrak{N}$ the coset x^- ($= x + \mathfrak{M}$). We know already that T is an isomorphism between $\mathfrak{N}$ and $\mathcal{V}/\mathfrak{M}$ (cf. § 22, Theorem 1); we shall show now that the isomorphism carries the transformation C over to the transformation A^-. If $Cx = y$ (where, of course, x is in $\mathfrak{N}$), then $A^- x^- = (Ax)^- = (Cx)^- = y^-$; it follows that $TCx = Ty = A^- Tx$. This implies that $TC = A^- T$, as promised. Loosely speaking (see § 47) we may say that A^- transforms $\mathcal{V}^-$ the same way as C transforms $\mathfrak{N}$. In other words, the linear transformations A^- and C are abstractly identical (isomorphic). This fact is of great significance in the applications of the concept of quotient space.

§ 49. Range and null-space

DEFINITION. If A is a linear transformation on a vector space $\mathcal{V}$ and if $\mathfrak{M}$ is a subspace of $\mathcal{V}$, the *image* of $\mathfrak{M}$ under A, in symbols $A\mathfrak{M}$, is the set of all vectors of the form Ax with x in $\mathfrak{M}$. The *range* of A is the set $\mathfrak{R}(A) = A\mathcal{V}$; the *null-space* of A is the set $\mathfrak{N}(A)$ of all vectors x for which $Ax = 0$.

It is immediately verified that $A\mathfrak{M}$ and $\mathfrak{N}(A)$ are subspaces. If, as usual, we denote by ϑ the subspace containing the vector 0 only, it is easy to describe some familiar concepts in terms of the terminology just introduced; we list some of the results.

(i) The transformation A is invertible if and only if $\mathfrak{R}(A) = \mathfrak{V}$ and $\mathfrak{N}(A) = \mathfrak{O}$.

(ii) In case $\mathfrak{V}$ is finite-dimensional, A is invertible if and only if $\mathfrak{R}(A) = \mathfrak{V}$ or $\mathfrak{N}(A) = \mathfrak{O}$.

(iii) The subspace $\mathfrak{M}$ is invariant under A if and only if $A\mathfrak{M} \subset \mathfrak{M}$.

(iv) A pair of complementary subspaces $\mathfrak{M}$ and $\mathfrak{N}$ reduce A if and only if $A\mathfrak{M} \subset \mathfrak{M}$ and $A\mathfrak{N} \subset \mathfrak{N}$.

(v) If E is the projection on $\mathfrak{M}$ along $\mathfrak{N}$, then $\mathfrak{R}(E) = \mathfrak{M}$ and $\mathfrak{N}(E) = \mathfrak{N}$.

All these statements are easy to prove; we indicate the proof of (v). From § 41, Theorem 2, we know that $\mathfrak{N}$ is the set of all solutions of the equation $Ex = 0$; this coincides with our definition of $\mathfrak{N}(E)$. We know also that $\mathfrak{M}$ is the set of all solutions of the equation $Ex = x$. If x is in $\mathfrak{M}$, then x is also in $\mathfrak{R}(E)$, since x is the image under E of something (namely of x itself). Conversely, if a vector x is the image under E of something, say, $x = Ey$ (so that x is in $\mathfrak{R}(E)$), then $Ex = E^2x = Ey = x$, so that x is in $\mathfrak{M}$.

Warning: it is accidental that for projections $\mathfrak{R} \oplus \mathfrak{N} = \mathfrak{V}$. In general it need not even be true that $\mathfrak{R} = \mathfrak{R}(A)$ and $\mathfrak{N} = \mathfrak{N}(A)$ are disjoint. It can happen, for example, that for a certain vector x we have $x \neq 0$, $Ax \neq 0$, and $A^2x = 0$; for such a vector, Ax clearly belongs to both the range and the null-space of A.

THEOREM. *If A is a linear transformation on a vector space $\mathfrak{V}$, then*

$$(1) \qquad\qquad (\mathfrak{R}(A))^0 = \mathfrak{N}(A');$$

if $\mathfrak{V}$ is finite-dimensional, then

$$(2) \qquad\qquad (\mathfrak{N}(A))^0 = \mathfrak{R}(A').$$

PROOF. If y is in $(\mathfrak{R}(A))^0$, then, for all x in $\mathfrak{V}$,

$$0 = [Ax, y] = [x, A'y],$$

so that $A'y = 0$ and y is in $\mathfrak{N}(A')$. If, on the other hand, y is in $\mathfrak{N}(A')$, then, for all x in $\mathfrak{V}$,

$$0 = [x, A'y] = [Ax, y],$$

so that y is in $(\mathfrak{R}(A))^0$.

If we apply (1) to A' in place of A, we obtain

$$(3) \qquad\qquad (\mathfrak{R}(A'))^0 = \mathfrak{N}(A'').$$

If $\mathfrak{V}$ is finite-dimensional (and hence reflexive), we may replace A'' by A in (3), and then we may form the annihilator of both sides; the desired conclusion (2) follows from § 17, Theorem 2.

1. Use the differentiation operator on $\mathcal{O}_n$ to show that the range and the null-space of a linear transformation need not be disjoint.

2. (a) Give an example of a linear transformation on a three-dimensional space with a two-dimensional range.

(b) Give an example of a linear transformation on a three-dimensional space with a two-dimensional null-space.

3. Find a four-by-four matrix whose range is spanned by $(1, 0, 1, 0)$ and $(0, 1, 0, 1)$.

4. (a) Two projections E and F have the same range if and only if $EF = F$ and $FE = E$.

(b) Two projections E and F have the same null-space if and only if $EF = E$ and $FE = F$.

5. If $E_1, \cdots, E_k$ are projections with the same range and if $\alpha_1, \cdots, \alpha_k$ are scalars such that $\sum_i \alpha_i = 1$, then $\sum_i \alpha_i E_i$ is a projection.

§ 50. Rank and nullity

We shall now restrict attention to the finite-dimensional case and draw certain easy conclusions from the theorem of the preceding section.

DEFINITION. The *rank*, $\rho(A)$, of a linear transformation A on a finite-dimensional vector space is the dimension of $\Re(A)$; the *nullity*, $\nu(A)$, is the dimension of $\Re(A)$.

THEOREM 1. *If A is a linear transformation on an n-dimensional vector space, then $\rho(A) = \rho(A')$ and $\nu(A) = n - \rho(A)$.*

PROOF. The theorem of the preceding section and § 17, Theorem 1, together imply that

$$(1) \qquad \nu(A') = n - \rho(A).$$

Let $\mathfrak{X} = \{x_1, \cdots, x_n\}$ be any basis for which $x_1, \cdots, x_\nu$ are in $\Re(A)$; then, for any $x = \sum_i \xi_i x_i$, we have

$$Ax = \sum_i \xi_i A x_i = \sum_{i=\nu+1}^n \xi_i A x_i.$$

In other words, Ax is a linear combination of the $n - \nu$ vectors $Ax_{\nu+1}$, $\cdots$, Ax_n; it follows that $\rho(A) \leqq n - \nu(A)$. Applying this result to A' and using (1), we obtain

$$(2) \qquad \rho(A') \leqq n - \nu(A') = \rho(A).$$

In (2) we may replace A by A', obtaining

(3) $$\rho(A) = \rho(A'') \leqq \rho(A');$$

(2) and (3) together show that

(4) $$\rho(A) = \rho(A'),$$

and (1) and (4) together show that

(5) $$\nu(A') = n - \rho(A').$$

Replacing A by A' in (5) gives, finally,

(6) $$\nu(A) = n - \rho(A),$$

and concludes the proof of the theorem.

These results are usually discussed from a little different point of view. Let A be a linear transformation on an n-dimensional vector space, and let $\mathfrak{X} = \{x_1, \cdots, x_n\}$ be a basis in that space; let $[A] = (\alpha_{ij})$ be the matrix of A in the coordinate system $\mathfrak{X}$, so that

$$Ax_j = \sum_i \alpha_{ij} x_i.$$

Since if $x = \sum_j \xi_j x_j$, then $Ax = \sum_j \xi_j A x_j$, it follows that every vector in $\mathfrak{R}(A)$ is a linear combination of the Ax_j, and hence of any maximal linearly independent subset of the Ax_j. It follows that the maximal number of linearly independent Ax_j is precisely $\rho(A)$. In terms of the co-ordinates $(\alpha_{1j}, \cdots, \alpha_{nj})$ of Ax_j we may express this by saying that $\rho(A)$ is the maximal number of linearly independent columns of the matrix $[A]$. Since (§ 45) the columns of $[A']$ (the matrix being expressed in terms of the dual basis of $\mathfrak{X}$) are the rows of $[A]$, it follows from Theorem 1 that $\rho(A)$ is also the maximal number of linearly independent rows of $[A]$. Hence "the row rank of $[A]$ = the column rank of $[A]$ = the rank of $[A]$."

THEOREM 2. *If A is a linear transformation on the n-dimensional vector space $\mathcal{V}$, and if $\mathfrak{K}$ is any h-dimensional subspace of $\mathcal{V}$, then the dimension of $A\mathfrak{K}$ is $\geqq h - \nu(A)$.*

PROOF. Let $\mathfrak{K}$ be any subspace for which $\mathcal{V} = \mathfrak{K} \oplus \mathfrak{K}$, so that if k is the dimension of $\mathfrak{K}$, then $k = n - h$. Upon operating with A we obtain

$$A\mathcal{V} = A\mathfrak{K} + A\mathfrak{K}.$$

(The sum is not necessarily a direct sum; see § 11.) Since $A\mathcal{V} = \mathfrak{R}(A)$ has dimension $n - \nu(A)$, since the dimension of $A\mathfrak{K}$ is clearly $\leqq k = n - h$, and since the dimension of the sum is $\leqq$ the sum of the dimensions, we have the desired result.

THEOREM 3. *If A and B are linear transformations on a finite-dimensional vector space, then*

(7) $$\rho(A + B) \leq \rho(A) + \rho(B),$$

(8) $$\rho(AB) \leq \min \{\rho(A), \rho(B)\},$$

and

(9) $$\nu(AB) \leq \nu(A) + \nu(B).$$

If B is invertible, then

(10) $$\rho(AB) = \rho(BA) = \rho(A).$$

PROOF. Since $(AB)x = A(Bx)$, it follows that $\Re(AB)$ is contained in $\Re(A)$, so that $\rho(AB) \leq \rho(A)$, or, in other words, the rank of a product is not greater than the rank of the first factor. Let us apply this auxiliary result to $B'A'$; this, together with what we already know, yields (8). If B is invertible, then

$$\rho(A) = \rho(AB \cdot B^{-1}) \leq \rho(AB)$$

and

$$\rho(A) = \rho(B^{-1} \cdot BA) \leq \rho(BA);$$

together with (8) this yields (10). The equation (7) is an immediate consequence of an argument we have already used in the proof of Theorem 2. The proof of (9) we leave as an exercise for the reader. (Hint: apply Theorem 2 with $\mathfrak{K} = B\mathfrak{V} = \Re(B)$.) Together the two formulas (8) and (9) are known as Sylvester's *law of nullity*.

§ 51. Transformations of rank one

We conclude our discussion of rank by a description of the matrices of linear transformations of rank ≤ 1.

THEOREM 1. *If a linear transformation A on a finite-dimensional vector space $\mathfrak{V}$ is such that $\rho(A) \leq 1$ (that is, $\rho(A) = 0$ or $\rho(A) = 1$), then the elements of the matrix $[A] = (\alpha_{ij})$ of A have the form $\alpha_{ij} = \beta_i \gamma_j$ in every coordinate system; conversely if the matrix of A has this form in some one coordinate system, then $\rho(A) \leq 1$.*

PROOF. If $\rho(A) = 0$, then $A = 0$, and the statement is trivial. If $\rho(A) = 1$, that is, $\Re(A)$ is one-dimensional, then there exists in $\Re(A)$ a non-zero vector x_0 (a basis in $\Re(A)$) such that every vector in $\Re(A)$ is a multiple of x_0. Hence, for every x,

$$Ax = y_0 x_0,$$

where the scalar coefficient $y_0(= y_0(x))$ depends, of course, on x. The linearity of A implies that y_0 is a linear functional on $\mathcal{V}$. Let $\mathfrak{X} = \{x_1, \cdots, x_n\}$ be a basis in $\mathcal{V}$, and let (α_{ij}) be the corresponding matrix of A, so that

$$Ax_j = \sum_i \alpha_{ij} x_i.$$

If $\mathfrak{X}' = \{y_1, \cdots, y_n\}$ is the dual basis in $\mathcal{V}'$, then (cf. § 45, (2))

$$\alpha_{ij} = [Ax_j, y_i].$$

In the present case

$$\alpha_{ij} = [y_0(x_j)x_0, y_i] = y_0(x_j)[x_0, y_i] = [x_0, y_i][x_j, y_0];$$

in other words, we may take $\beta_i = [x_0, y_i]$ and $\gamma_j = [x_j, y_0]$.

Conversely, suppose that in a fixed coordinate system $\mathfrak{X} = \{x_1, \cdots, x_n\}$ the matrix (α_{ij}) of A is such that $\alpha_{ij} = \beta_i\gamma_j$. We may find a linear functional y_0 such that $\gamma_j = [x_j, y_0]$, and we may define a vector x_0 by $x_0 = \sum_k \beta_k x_k$. The linear transformation $\tilde{A}$ defined by $\tilde{A}x = y_0(x)x_0$ is clearly of rank one (unless, of course, $\alpha_{ij} = 0$ for all i and j), and its matrix $(\tilde{\alpha}_{ij})$ in the coordinate system $\mathfrak{X}$ is given by

$$\tilde{\alpha}_{ij} = [\tilde{A}x_j, y_i]$$

(where $\mathfrak{X}' = \{y_1, \cdots, y_n\}$ is the dual basis of $\mathfrak{X}$). Hence

$$\tilde{\alpha}_{ij} = [y_0(x_j)x_0, y_i] = [x_0, y_i][x_j, y_0] = \beta_i\gamma_j,$$

and, since A and $\tilde{A}$ have the same matrix in one coordinate system, it follows that $\tilde{A} = A$. This concludes the proof of the theorem.

The following theorem sometimes makes it possible to apply Theorem 1 to obtain results about an arbitrary linear transformation.

THEOREM 2. *If A is a linear transformation of rank ρ on a finite-dimensional vector space $\mathcal{V}$, then A may be written as the sum of ρ transformations of rank one.*

PROOF. Since $A\mathcal{V} = \mathcal{R}(A)$ has dimension ρ, we may find ρ vectors $x_1, \cdots, x_\rho$ that form a basis for $\mathcal{R}(A)$. It follows that, for every vector x in $\mathcal{V}$, we have

$$Ax = \sum_{i=1}^{\rho} \xi_i x_i,$$

where each ξ_i depends, of course, on x; we write $\xi_i = y_i(x)$. It is easy to see that y_i is a linear functional. In terms of these y_i we define, for each $i = 1, \cdots, \rho$, a linear transformation A_i by $A_i x = y_i(x)x_i$. It follows that each A_i has rank one and $A = \sum_{i=1}^{\rho} A_i$. (Compare this result with § 32, example (2).)

A slight refinement of the proof just given yields the following result.

THEOREM 3. *Corresponding to any linear transformation A on a finite-dimensional vector space $\mathcal{V}$ there is an invertible linear transformation P for which PA is a projection.*

PROOF. Let $\mathcal{R}$ and $\mathcal{N}$, respectively, be the range and the null-space of A, and let $\{x_1, \cdots, x_\rho\}$ be a basis for $\mathcal{R}$. Let $x_{\rho+1}, \cdots, x_n$ be vectors such that $\{x_1, \cdots, x_n\}$ is a basis for $\mathcal{V}$. Since x_i is in $\mathcal{R}$ for $i = 1, \cdots, \rho$, we may find vectors y_i such that $Ay_i = x_i$; finally, we choose a basis for $\mathcal{N}$, which we may denote by $\{y_{\rho+1}, \cdots, y_n\}$. We assert that $\{y_1, \cdots, y_n\}$ is a basis for $\mathcal{V}$. We need, of course, to prove only that the y's are linearly independent. For this purpose we suppose that $\sum_{i=1}^n \alpha_i y_i = 0$; then we have (remembering that for $i = \rho + 1, \cdots, n$ the vector y_i belongs to $\mathcal{N}$)

$$A\left(\sum_{i=1}^n \alpha_i y_i\right) = \sum_{i=1}^\rho \alpha_i x_i = 0,$$

whence $\alpha_1 = \cdots = \alpha_\rho = 0$. Consequently $\sum_{i=\rho+1}^n \alpha_i y_i = 0$; the linear independence of $y_{\rho+1}, \cdots, y_n$ shows that the remaining α's must also vanish.

A linear transformation P, of the kind whose existence we asserted, is now determined by the conditions $Px_i = y_i$, $i = 1, \cdots, n$. Indeed, if $i = 1, \cdots, \rho$, then $PAy_i = Px_i = y_i$, and if $i = \rho + 1, \cdots, n$, then $PAy_i = P0 = 0$.

Consideration of the adjoint of A, together with the reflexivity of $\mathcal{V}$, shows that we may also find an invertible Q for which AQ is a projection. In case A itself is invertible, we must have $P = Q = A^{-1}$.

EXERCISES

1. What is the rank of the differentiation operator on $\mathcal{P}_n$? What is its nullity?

2. Find the ranks of the following matrices.

(a) $\begin{pmatrix} 1 & 1 & 1 \\ 1 & 1 & 1 \\ 1 & 1 & 1 \end{pmatrix}$.

(c) $\begin{pmatrix} 0 & 0 & 1 \\ 0 & 1 & 0 \\ 1 & 0 & 0 \end{pmatrix}$.

(b) $\begin{pmatrix} 1 & 1 & 1 \\ 1 & 1 & 0 \\ 1 & 0 & 0 \end{pmatrix}$.

(d) $\begin{pmatrix} 0 & 1 & 0 \\ 1 & 0 & 1 \\ 0 & 1 & 0 \end{pmatrix}$.

3. If A is left multiplication by P on a space of linear transformations (cf. § 38, Ex. 5), and if P has rank m, what is the rank of A?

4. The rank of the direct sum of two linear transformations (on finite-dimensional vector spaces) is the sum of their ranks.

5. (a) If A and B are linear transformations on an n-dimensional vector space, and if $AB = 0$, then $\rho(A) + \rho(B) \leq n$.

(b) For each linear transformation A on an n-dimensional vector space there exists a linear transformation B such that $AB = 0$ and such that $\rho(A) + \rho(B) = n$.

6. If A, B, and C are linear transformations on a finite-dimensional vector space, then

$$\rho(AB) + \rho(BC) \leq \rho(B) + \rho(ABC).$$

7. Prove that two linear transformations (on the same finite-dimensional vector space) are equivalent if and only if they have the same rank.

8. (a) Suppose that A and B are linear transformations (on the same finite-dimensional vector space) such that $A^2 = A$ and $B^2 = B$. Is it true that A and B are similar if and only if $\rho(A) = \rho(B)$?
(b) Suppose that A and B are linear transformations (on the same finite-dimensional vector space) such that $A \neq 0$, $B \neq 0$, and $A^2 = B^2 = 0$. Is it true that A and B are similar if and only if $\rho(A) = \rho(B)$?

9. (a) If A is a linear transformation of rank one, then there exists a unique scalar α such that $A^2 = \alpha A$.
(b) If $\alpha \neq 1$, then $1 - A$ is invertible.

§ 52. Tensor products of transformations

Let us now tie up linear transformations with the theory of tensor products. Let $\mathfrak{U}$ and $\mathfrak{V}$ be finite-dimensional vector spaces (over the same field), and let A and B be any two linear transformations on $\mathfrak{U}$ and $\mathfrak{V}$ respectively. We define a linear transformation $\bar{C}$ on the space $\mathfrak{W}$ of all bilinear forms on $\mathfrak{U} \oplus \mathfrak{V}$ by writing

$$(\bar{C}w)(x, y) = w(Ax, By).$$

The *tensor product* $C = A \otimes B$ of the transformations A and B is, by definition, the dual of the transformation $\bar{C}$, so that

$$(Cz)(w) = z(\bar{C}w)$$

whenever z is in $\mathfrak{U} \otimes \mathfrak{V}$ and w is in $\mathfrak{W}$. If we apply C to an element z_0 of the form $z_0 = x_0 \otimes y_0$ (recall that this means that $z_0(w) = w(x_0, y_0)$ for all w in $\mathfrak{W}$), we obtain

$$(Cz_0)(w) = z_0(\bar{C}w) = (x_0 \otimes y_0)(\bar{C}w)$$

$$= (\bar{C}w)(x_0, y_0) = w(Ax_0, By_0) = (Ax_0 \otimes By_0)(w).$$

We infer that

(1) $$Cz_0 = Ax_0 \otimes By_0.$$

Since there are quite a few elements in $\mathfrak{U} \otimes \mathfrak{V}$ of the form $x \otimes y$, enough at any rate to form a basis (see § 25), this relation characterizes C.

The formal rules for operating with tensor products go as follows.

(2) $$A \otimes 0 = 0 \otimes B = 0,$$

(3) $$1 \otimes 1 = 1,$$

(4) $$(A_1 + A_2) \otimes B = (A_1 \otimes B) + (A_2 \otimes B),$$

(5) $$A \otimes (B_1 + B_2) = (A \otimes B_1) + (A \otimes B_2),$$

(6) $$\alpha A \otimes \beta B = \alpha\beta(A \otimes B),$$

(7) $$(A \otimes B)^{-1} = A^{-1} \otimes B^{-1},$$

(8) $$(A_1 A_2) \otimes (B_1 B_2) = (A_1 \otimes B_1)(A_2 \otimes B_2).$$

The proofs of all these relations, except perhaps the last two, are straightforward.

Formula (7), as all formulas involving inverses, has to be read with caution. It is intended to mean that if both A and B are invertible, then so is $A \otimes B$, and the equation holds, and, conversely, that if $A \otimes B$ is invertible, then so also are A and B. We shall prove (7) and (8) in reverse order.

Formula (8) follows from the characterization (1) of tensor products and the following computation:

$$(A_1 A_2 \otimes B_1 B_2)(x \otimes y) = A_1 A_2 x \otimes B_1 B_2 y$$

$$= (A_1 \otimes B_1)(A_2 x \otimes B_2 y) = (A_1 \otimes B_1)(A_2 \otimes B_2)(x \otimes y).$$

As an immediate consequence of (8) we obtain

(9) $$A \otimes B = (A \otimes 1)(1 \otimes B) = (1 \otimes B)(A \otimes 1).$$

To prove (7), suppose that A and B are invertible, and form $A \otimes B$ and $A^{-1} \otimes B^{-1}$. Since, by (8), the product of these two transformations, in either order, is 1, it follows that $A \otimes B$ is invertible and that (7) holds. Conversely, suppose that $A \otimes B$ is invertible. Remembering that we defined tensor products for finite-dimensional spaces only, we may invoke § 36, Theorem 2; it is sufficient to prove that $Ax = 0$ implies that $x = 0$ and $By = 0$ implies that $y = 0$. We use (1):

$$Ax \otimes By = (A \otimes B)(x \otimes y).$$

If either factor on the left is zero, then $(A \otimes B)(x \otimes y) = 0$, whence $x \otimes y = 0$, so that either $x = 0$ or $y = 0$. Since (by (2)) $B = 0$ is impossible, we may find a vector y so that $By \neq 0$. Applying the above argument to this y, with any x for which $Ax = 0$, we conclude that $x = 0$. The same argument with the roles of A and B interchanged proves that B is invertible.

An interesting (and complicated) side of the theory of tensor products of transformations is the theory of Kronecker products of matrices. Let $\mathfrak{X} = \{x_1, \cdots, x_n\}$ and $\mathcal{Y} = \{y_1, \cdots, y_m\}$ be bases in $\mathfrak{U}$ and $\mathfrak{V}$, and let $[A] = [A; \mathfrak{X}] = (\alpha_{ij})$ and $[B] = [B; \mathcal{Y}] = (\beta_{pq})$ be the matrices of A and B. What is the matrix of $A \otimes B$ in the coordinate system $\{x_i \otimes y_p\}$?

To answer the question, we must recall the discussion in § 37 concerning the arrangement of a basis in a linear order. Since, unfortunately, it is impossible to write down a matrix without being committed to an order of the rows and the columns, we shall be frank about it, and arrange the n times m vectors $x_i \otimes y_p$ in the so-called lexicographical order, as follows:

$$x_1 \otimes y_1, x_1 \otimes y_2, \cdots, x_1 \otimes y_m, x_2 \otimes y_1, \cdots,$$

$$x_2 \otimes y_m, \cdots, x_n \otimes y_1, \cdots, x_n \otimes y_m.$$

We proceed also to carry out the following computation:

$$(A \otimes B)(x_j \otimes y_q) = Ax_j \otimes By_q = \left(\sum_i \alpha_{ij}x_i\right) \otimes \left(\sum_p \beta_{pq}y_p\right)$$

$$= \sum_i \sum_p \alpha_{ij}\beta_{pq}(x_i \otimes y_p).$$

This process indicates exactly how far we can get without ordering the basis elements; if, for example, we agree to index the elements of a matrix not with a pair of integers but with a pair of pairs, say (i, p) and (j, q), then we know now that the element in the (i, p) row and the (j, q) column is $\alpha_{ij}\beta_{pq}$. If we use the lexicographical ordering, the matrix of $A \otimes B$ has the form

$$\begin{bmatrix} \alpha_{11}\beta_{11} \cdots \alpha_{11}\beta_{1m} & \cdots & \alpha_{1n}\beta_{11} \cdots \alpha_{1n}\beta_{1m} \\ \vdots & & \vdots \\ \alpha_{11}\beta_{m1} \cdots \alpha_{11}\beta_{mm} & \cdots & \alpha_{1n}\beta_{m1} \cdots \alpha_{1n}\beta_{mm} \\ \vdots & & \vdots \\ \alpha_{n1}\beta_{11} \cdots \alpha_{n1}\beta_{1m} & \cdots & \alpha_{nn}\beta_{11} \cdots \alpha_{nn}\beta_{1m} \\ \vdots & & \vdots \\ \alpha_{n1}\beta_{m1} \cdots \alpha_{n1}\beta_{mm} & \cdots & \alpha_{nn}\beta_{m1} \cdots \alpha_{nn}\beta_{mm} \end{bmatrix}.$$

In a condensed notation whose meaning is clear we may write this matrix as

$$\begin{bmatrix} \alpha_{11}[B] & \cdots & \alpha_{1n}[B] \\ \cdot & & \cdot \\ \cdot & & \cdot \\ \cdot & & \cdot \\ \alpha_{n1}[B] & \cdots & \alpha_{nn}[B] \end{bmatrix}.$$

This matrix is known as the *Kronecker product* of $[A]$ and $[B]$, in that order. The rule for forming it is easy to describe in words: replace each element α_{ij} of the n-by-n matrix $[A]$ by the m-by-m matrix $\alpha_{ij}[B]$. If in this rule we interchange the roles of A and B (and consequently interchange n and m) we obtain the definition of the Kronecker product of $[B]$ and $[A]$.

EXERCISES

1. We know that the tensor product of $\mathcal{O}_n$ and $\mathcal{O}_m$ may be identified with the space $\mathcal{O}_{n,m}$ of polynomials in two variables (see § 25, Ex. 2). Prove that if A and B are differentiation on $\mathcal{O}_n$ and $\mathcal{O}_m$ respectively, and if $C = A \otimes B$, then C is mixed partial differentiation, that is, if z is in $\mathcal{O}_{n,m}$, then $Cz = \dfrac{\partial^2 z}{\partial s\, \partial t}$.

2. With the lexicographic ordering of the product basis $\{x_i \otimes y_p\}$ it turned out that the matrix of $A \otimes B$ is the Kronecker product of the matrices of A and B. Is there an arrangement of the basis vectors such that the matrix of $A \otimes B$, referred to the coordinate system so arranged, is the Kronecker product of the matrices of B and A (in that order)?

3. If A and B are linear transformations, then

$$\rho(A \otimes B) = \rho(A)\rho(B).$$

§ 53. Determinants

It is, of course, possible to generalize the considerations of the preceding section to multilinear forms and multiple tensor products. Instead of entering into that part of multilinear algebra, we proceed in a different direction; we go directly after determinants.

Suppose that A is a linear transformation on an n-dimensional vector space $\mathcal{V}$ and let w be an alternating n-linear form on $\mathcal{V}$. If we write $\bar{A}w$ for the function defined by

$$(\bar{A}w)(x_1, \cdots, x_n) = w(Ax_1, \cdots, Ax_n),$$

then $\bar{A}w$ is an alternating n-linear form on $\mathcal{V}$, and, in fact, $\bar{A}$ is a linear transformation on the space of such forms. Since (see § 31) that space is one-dimensional, it follows that $\bar{A}$ is equal to multiplication by an appropriate scalar. In other words, there exists a scalar δ such that $\bar{A}w = \delta w$ for every alternating n-linear form w. By this somewhat roundabout procedure (from A to $\bar{A}$ to δ) we have associated a uniquely determined scalar δ with every linear transformation A on $\mathbb{V}$; we call δ the *determinant* of A, and we write $\delta = \det A$. Observe that *det* is neither a scalar nor a transformation, but a function that associates a scalar with each linear transformation.

Our immediate purpose is to study the function *det*. We begin by finding the determinants of the simplest linear transformations, that is, the multiplications by scalars. If $Ax = \alpha x$ for every x in $\mathbb{V}$, then

$$(\bar{A}w)(x_1, \cdots, x_n) = w(\alpha x_1, \cdots, \alpha x_n) = \alpha^n w(x_1, \cdots, x_n)$$

for every alternating n-linear form w; it follows that $\det A = \alpha^n$. We note, in particular, that $\det 0 = 0$ and $\det 1 = 1$.

Next we ask about the multiplicative properties of *det*. Suppose that A and B are linear transformations on $\mathbb{V}$, and write $C = AB$. If w is an alternating n-linear form, then

$$(\bar{C}w)(x_1, \cdots, x_n) = w(ABx_1, \cdots, ABx_n)$$
$$= (\bar{A}w)(Bx_1, \cdots, Bx_n) = (\bar{B}\bar{A}w)(x_1, \cdots, x_n),$$

so that $\bar{C} = \bar{B}\bar{A}$. Since

$$\bar{C}w = (\det C)w$$

and

$$\bar{B}\bar{A}w = (\det B)\bar{A}w = (\det B)(\det A)w,$$

it follows that

$$\det(AB) = (\det A)(\det B).$$

(The values of *det* are scalars, and therefore commute with each other.)

A linear transformation A is called *singular* if $\det A = 0$ and *non-singular* otherwise. Our next result is that A is invertible if and only if it is non-singular. Indeed, if A is invertible, then

$$1 = \det 1 = \det(AA^{-1}) = (\det A)(\det A^{-1}),$$

and therefore $\det A \neq 0$. Suppose, on the other hand, that $\det A \neq 0$. If $\{x_1, \cdots, x_n\}$ is a basis in $\mathbb{V}$, and if w is a non-zero alternating n-linear form on $\mathbb{V}$, then $(\det A)w(x_1, \cdots, x_n) \neq 0$ by § 30, Theorem 3. This implies, by § 30, Theorem 2, that the set $\{Ax_1, \cdots, Ax_n\}$ is linearly independent (and therefore a basis); from this, in turn, we infer that A is invertible.

In the classical literature determinant is defined as a function of matrices (not linear transformations); we are now in a position to make contact with that approach. We shall derive an expression for det A in terms of the elements α_{ij} of the matrix corresponding to A in some coordinate system $\{x_1, \cdots, x_n\}$. Let w be a non-zero alternating n-linear form; we know that

$$(1) \qquad (\det A)w(x_1, \cdots, x_n) = w(Ax_1, \cdots, Ax_n).$$

If we replace each Ax_j in the right side of (1) by $\sum_i \alpha_{ij}x_i$ and expand the result by multilinearity, we obtain a long linear combination of terms such as $w(z_1, \cdots, z_n)$, where each z is one of the x's. (Compare this part of the argument with the proof of § 30, Theorem 3.) If, in such a term, two of the z's coincide, then, since w is alternating, that term must vanish. If, on the other hand, all the z's are distinct, then $w(z_1, \cdots, z_n) = \pi w(x_1, \cdots, x_n)$ for some permutation π, and, moreover, every permutation π can occur in this way. The coefficient of the term $\pi w(x_1, \cdots, x_n)$ is the product $\alpha_{\pi(1),1} \cdots \alpha_{\pi(n),n}$. Since (§ 30, Theorem 1) w is skew symmetric, it follows that

$$(2) \qquad \det A = \sum_\pi (\operatorname{sgn} \pi)\alpha_{\pi(1),1} \cdots \alpha_{\pi(n),n}$$

where the summation is extended over all permutations π in S_n. (Recall that $w(x_1, \cdots, x_n) \neq 0$, by § 30, Theorem 3, so that division by $w(x_1, \cdots, x_n)$ is legitimate.)

From this classical equation (2) we could derive many special properties of determinants by straightforward computation. Here is one example. If σ and π are permutations (in S_n), then (since $\pi\sigma$ is also a permutation), it follows that the products $\alpha_{\pi(1),1} \cdots \alpha_{\pi(n),n}$ and $\alpha_{\pi\sigma(1),\sigma(1)} \cdots \alpha_{\pi\sigma(n),\sigma(n)}$ differ in the order of their factors only. If, for each π, we take $\sigma = \pi^{-1}$, and then alter each summand in (2) accordingly, we obtain

$$\det A = \sum_\pi (\operatorname{sgn} \pi)\alpha_{1,\pi(1)} \cdots \alpha_{n,\pi(n)}.$$

(Note that $\operatorname{sgn} \pi = \operatorname{sgn} \pi^{-1}$ and that the sum over all π is the same as the sum over all π^{-1}.) Since this last sum is just like the sum in (2), except that $\alpha_{i,\pi(i)}$ appears in place of $\alpha_{\pi(i),i}$, it follows from an application of (2) to A' in place of A that

$$\det A = \det A'.$$

Here is another useful fact about determinants. If $\mathfrak{M}$ is a subspace invariant under A, if B is the transformation A considered on $\mathfrak{M}$ only, and if C is the quotient transformation $A/\mathfrak{M}$, then

$$\det A = \det B \cdot \det C.$$

This multiplicative relation holds if, in particular, A is the direct sum of two transformations B and C. The proof can be based directly on the definition of determinants, or, alternatively, on the expansion obtained in the preceding paragraph.

If, for a fixed linear transformation A, we write $p(\lambda) = \det (A - \lambda)$, then p is a function of the scalar λ; we assert that it is, in fact, a polynomial of degree n in λ, and that the coefficient of λ^n is $(-1)^n$. For the proof we may use the notation of (1). It is easy to see that $w((A - \lambda)x_1, \cdots, (A - \lambda)x_n)$ is a sum of terms such as $\lambda^k w(y_1, \cdots, y_n)$, where $y_i = x_i$ for exactly k values of i and $y_i = Ax_i$ for the remaining $n - k$ values of i $(k = 0, 1, \cdots, n)$. The polynomial p is called the *characteristic polynomial* of A; the equation $p = 0$, that is, $\det (A - \lambda) = 0$, is the *characteristic equation* of A. The roots of the characteristic equation of A (that is, the scalars α such that $\det (A - \alpha) = 0$) are called the *characteristic roots* of A.

EXERCISES

1. Use determinants to get a new proof of the fact that if A and B are linear transformations on a finite-dimensional vector space, and if $AB = 1$, then both A and B are invertible.

2. If A and B are linear transformations such that $AB = 0$, $A \neq 0$, $B \neq 0$, then $\det A = \det B = 0$.

3. Suppose that (α_{ij}) is a non-singular n-by-n matrix, and suppose that $A_1, \cdots, A_n$ are linear transformations (on the same vector space). Prove that if the linear transformations $\sum_j \alpha_{ij} A_j$, $i = 1, \cdots, n$, commute with each other, then the same is true of $A_1, \cdots, A_n$.

4. If $\{x_1, \cdots, x_n\}$ and $\{y_1, \cdots, y_n\}$ are bases in the same vector space, and if A is a linear transformation such that $Ax_i = y_i$, $i = 1, \cdots, n$, then $\det A \neq 0$.

5. Suppose that $\{x_1, \cdots, x_n\}$ is a basis in a finite-dimensional vector space $\mathcal{V}$. If $y_1, \cdots, y_n$ are vectors in $\mathcal{V}$, write $w(y_1, \cdots, y_n)$ for the determinant of the linear transformation A such that $Ax_j = y_j$, $j = 1, \cdots, n$. Prove that w is an alternating n-linear form.

6. If, in accordance with § 53, (2), the determinant of a matrix (α_{ij}) (not a linear transformation) is defined to be $\sum_\pi (\text{sgn } \pi)\alpha_{\pi(1),1} \cdots \alpha_{\pi(n),n}$, then, for each linear transformation A, the determinants of all the matrices $[A; \mathfrak{X}]$ are all equal to each other. (Here $\mathfrak{X}$ is an arbitrary basis.)

7. If (α_{ij}) is an n-by-n matrix such that $\alpha_{ij} = 0$ for more than $n^2 - n$ pairs of values of i and j, then $\det (\alpha_{ij}) = 0$.

8. If A and B are linear transformations on vector spaces of dimensions n and m, respectively, then

$$\det (A \otimes B) = (\det A)^m \cdot (\det B)^n.$$

9. If A, B, C, and D are matrices such that C and D commute and D is invertible, then (cf. § 38, Ex. 19)

$$\det \begin{pmatrix} A & B \\ C & D \end{pmatrix} = \det (AD - BC).$$

(Hint: multiply on the right by $\begin{pmatrix} 1 & 0 \\ X & 1 \end{pmatrix}$.) What if D is not invertible? What if C and D do not commute?

10. Do A and A' always have the same characteristic polynomial?

11. (a) If A and B are similar, then $\det A = \det B$.
(b) If A and B are similar, then A and B have the same characteristic polynomial.
(c) If A and B have the same characteristic polynomial, then $\det A = \det B$.
(d) Is the converse of any of these assertions true?

12. Determine the characteristic polynomial of the matrix (or, rather, of the linear transformation defined by the matrix)

$$\begin{bmatrix} 0 & 1 & 0 & \cdots & 0 \\ 0 & 0 & 1 & \cdots & 0 \\ \cdot & \cdot & \cdot & & \cdot \\ \cdot & \cdot & \cdot & & \cdot \\ \cdot & \cdot & \cdot & & \cdot \\ 0 & 0 & 0 & \cdots & 1 \\ \alpha_{n-1} & \alpha_{n-2} & \alpha_{n-3} & \cdots & \alpha_0 \end{bmatrix},$$

and conclude that every polynomial is the characteristic polynomial of some linear transformation.

13. Suppose that A and B are linear transformations on the same finite-dimensional vector space.
(a) Prove that if A is a projection, then AB and BA have the same characteristic polynomial. (Hint: choose a basis that makes the matrix of A as simple as possible and then compute directly with matrices.)
(b) Prove that, in all cases, AB and BA have the same characteristic polynomial. (Hint: find an invertible P such that PA is a projection and apply (a) to PA and BP^{-1}.)

§ 54. Proper values

A scalar λ is a *proper value* and a non-zero vector x is a *proper vector* of a linear transformation A if $Ax = \lambda x$. Almost every combination of the adjectives proper, latent, characteristic, eigen, and secular, with the nouns root, number, and value, has been used in the literature for what we call a proper value. It is important to be aware of the order of choice in the definition; λ is a proper value of A if there exists a non-zero vector x for which $Ax = \lambda x$, and a non-zero vector x is a proper vector of A if there exists a scalar λ for which $Ax = \lambda x$.

Suppose that λ is a proper value of A; let $\mathfrak{M}$ be the collection of all vectors x that are proper vectors of A belonging to this proper value, that is, for which $Ax = \lambda x$. Since, by our definition, 0 is not a proper vector, $\mathfrak{M}$ does not contain 0; if, however, we enlarge $\mathfrak{M}$ by adjoining the origin to it, then $\mathfrak{M}$ becomes a subspace. We define the *multiplicity* of the proper value λ as the dimension of the subspace $\mathfrak{M}$; a *simple* proper value is one whose multiplicity is equal to 1. By an obvious extension of this terminology, we may express the fact that a scalar λ is not a proper value of A at all by saying that λ is a proper value of multiplicity zero. The set of proper values of A is sometimes called the *spectrum* of A. Note that the spectrum of A is the same as the set of all scalars λ for which $A - \lambda$ is not invertible.

If the vector space we are working with has dimension n, then the scalar 0 is a proper value of multiplicity n of the linear transformation 0, and, similarly, the scalar 1 is a proper value of multiplicity n of the linear transformation 1. Since $Ax = \lambda x$ if and only if $(A - \lambda)x = 0$, that is, if and only if x is in the null-space of $A - \lambda$, it follows that the multiplicity of λ as a proper value of A is the same as the nullity of the linear transformation $A - \lambda$. From this, in turn, we infer (see § 50, Theorem 1) that the proper values of A, together with their associated multiplicities, are exactly the same as those of A'.

We observe that if B is any invertible transformation, then

$$BAB^{-1} - \lambda = B(A - \lambda)B^{-1},$$

so that $(A - \lambda)x = 0$ if and only if $(BAB^{-1} - \lambda)Bx = 0$. This implies that all spectral concepts (for example, the spectrum and the multiplicities of the proper values) are invariant under the replacement of A by BAB^{-1}. We note also that if $Ax = \lambda x$, then

$$A^2 x = A(Ax) = A(\lambda x) = \lambda(Ax) = \lambda(\lambda x) = \lambda^2 x.$$

More generally, if p is any polynomial, then $p(A)x = p(\lambda)x$, so that every proper vector of A, belonging to the proper value λ, is also a proper vector of $p(A)$, belonging to the proper value $p(\lambda)$. Hence if A satisfies any equation of the form $p(A) = 0$, then $p(\lambda) = 0$ for every proper value λ of A.

Since a necessary and sufficient condition that $A - \lambda$ have a non-trivial null-space is that it be singular, that is, that $\det(A - \lambda) = 0$, it follows that λ is a proper value of A if and only if it is a characteristic root of A. This fact is the reason for the importance of determinants in linear algebra. The useful geometric concept is that of a proper value. From the geometry of the situation, however, it is impossible to prove that any proper values exist. By means of determinants we reduce the problem to an algebraic one; it turns out that proper values are the same as roots of a certain polynomial equation. No wonder now that it is hard to prove that proper values always exist: polynomial equations do not always have roots, and, correspondingly, there are easy examples of linear transformations with no proper values.

§ 55. Multiplicity

The discussion in the preceding section indicates one of our reasons for wanting to study complex vector spaces. By the so-called fundamental theorem of algebra, a polynomial equation over the field of complex numbers always has at least one root; it follows that a linear transformation on a complex vector space always has at least one proper value. There are other fields, besides the field of complex numbers, over which every polynomial equation is solvable; they are called *algebraically closed* fields. The most general result of the kind we are after at the moment is that every linear transformation on a finite-dimensional vector space over an algebraically closed field has at least one proper value. Throughout the rest of this chapter (in the next four sections) we shall assume that our field of scalars is algebraically closed. The use we shall make of this assumption is the one just mentioned, namely, that from it we may conclude that proper values always exist.

The algebraic point of view on proper values suggests another possible definition of multiplicity. Suppose that A is a linear transformation on a finite-dimensional vector space, and suppose that λ is a proper value of A. We might wish to consider the multiplicity of λ as a root of the characteristic equation of A. This is a useful concept, which we shall call the *algebraic* multiplicity of λ, to distinguish it from our earlier, *geometric*, notion of multiplicity.

The two concepts of multiplicity do not coincide, as the following example shows. If D is differentiation on the space $\mathcal{P}_n$ of all polynomials of degree $\leq n - 1$, then a necessary and sufficient condition that a vector x in $\mathcal{P}_n$ be a proper vector of D is that $\dfrac{dx}{dt} \equiv \lambda x(t)$ for some complex number λ.

We borrow from the elementary theory of differential equations the fact that every solution of this equation is a constant multiple of $e^{\lambda t}$. Since, unless $\lambda = 0$, only the zero multiple of $e^{\lambda t}$ is a polynomial (which it must be if it is to belong to $\mathcal{P}_n$), we must have $\lambda = 0$ and $x(t) = 1$. In other words, this particular transformation has only one proper value (which must therefore occur with algebraic multiplicity n), namely, $\lambda = 0$; but, and this is more disturbing, the dimension of the linear manifold of solutions is exactly one. Hence if $n > 1$, the two definitions of multiplicity give different values. (In this argument we used the simple fact that a polynomial equation of degree n over an algebraically closed field has exactly n roots, if multiplicities are suitably counted. It follows that a linear transformation on an n-dimensional vector space over such a field has exactly n proper values, counting algebraic multiplicities.)

It is quite easy to see that the geometric multiplicity of λ is never greater than its algebraic multiplicity. Indeed, if A is any linear transformation, if λ_0 is any of its proper values, and if $\mathfrak{M}$ is the subspace of solutions of $Ax = \lambda_0 x$, then it is clear that $\mathfrak{M}$ is invariant under A. If A_0 is the linear transformation A considered on $\mathfrak{M}$ only, then it is clear that $\det (A_0 - \lambda)$ is a factor of $\det (A - \lambda)$. If the dimension of $\mathfrak{M}$ (= the geometric multiplicity of λ_0) is m, then $\det (A_0 - \lambda) = (\lambda_0 - \lambda)^m$; the desired result follows from the definition of algebraic multiplicity. It follows also that if $\lambda_1, \cdots, \lambda_p$ are the *distinct* proper values of A, with respective geometric multiplicities $m_1, \cdots, m_p$, and if it happens that $\sum_{i=1}^{p} m_i = n$, then m_i is equal to the algebraic multiplicity of λ_i for each $i = 1, \cdots, p$.

By means of proper values and their algebraic multiplicities we can characterize two interesting functions of linear transformations; one of them is the determinant and the other is something new. (Warning: these characterizations are valid only under our current assumption that the scalar field is algebraically closed.)

Let A be any linear transformation on an n-dimensional vector space, and let $\lambda_1, \cdots, \lambda_p$ be its distinct proper values. Let us denote by m_j the algebraic multiplicity of $\lambda_j, j = 1, \cdots, p$, so that $m_1 + \cdots + m_p = n$. For any polynomial equation

$$\alpha_0 + \alpha_1 \lambda + \cdots + \alpha_n \lambda^n = 0,$$

the product of the roots is $(-1)^n \alpha_0/\alpha_n$ and the sum of the roots is $-\alpha_{n-1}/\alpha_n$. Since the leading coefficient $(= \alpha_n)$ of the characteristic polynomial $\det (A - \lambda)$ is $(-1)^n$ and since the constant term $(= \alpha_0)$ is $\det (A - 0) = \det A$, we have

$$\det A = \prod_{j=1}^{p} \lambda_j^{m_j}.$$

This characterization of the determinant motivates the definition

$$\operatorname{tr} A = \sum_{j=1}^{p} m_j \lambda_j;$$

the function so defined is called the *trace* of A. We shall have no occasion to use trace in the sequel; we leave the derivation of the basic properties of the trace to the interested reader.

1. Find all (complex) proper values and proper vectors of the following matrices.

(a) $\begin{pmatrix} 0 & 1 \\ 0 & 0 \end{pmatrix}$.

(b) $\begin{pmatrix} 1 & 0 \\ 0 & i \end{pmatrix}$.

(c) $\begin{pmatrix} 1 & 1 \\ 0 & i \end{pmatrix}$.

(d) $\begin{pmatrix} 1 & 1 & 1 \\ 1 & 1 & 1 \\ 1 & 1 & 1 \end{pmatrix}$.

(e) $\begin{pmatrix} 1 & 1 & 1 \\ 0 & 1 & 1 \\ 0 & 0 & 1 \end{pmatrix}$.

2. Let π be a permutation of the integers $\{1, \cdots, n\}$; if $x = (\xi_1, \cdots, \xi_n)$ is a vector in $\mathbb{C}^n$, write $Ax = (\xi_{\pi(1)}, \cdots, \xi_{\pi(n)})$. Find the spectrum of A.

3. Prove that all the proper values of a projection are 0 or 1 and that all the proper values of an involution are $+1$ or -1. (This result does not depend on the finite-dimensionality of the vector space.)

4. Suppose that A is a linear transformation and that p is a polynomial. We know that if λ is a proper value of A, then $p(\lambda)$ is a proper value of $p(A)$; what can be said about the converse?

5. Prove that the differentiation operator D on the space $\mathcal{P}_n$ ($n > 1$) is not reducible (that is, it is not reduced by any non-trivial pair of complementary subspaces $\mathfrak{M}$ and $\mathfrak{N}$).

6. If A is a linear transformation on a finite-dimensional vector space, and if λ is a proper value of A, then the algebraic multiplicity of λ for A is equal to the algebraic multiplicity of λ for BAB^{-1}. (Here B is an arbitrary invertible transformation.)

7. Do AB and BA always have the same spectrum?

8. Suppose that A and B are linear transformations on finite-dimensional vector spaces.
   (a) $\mathrm{tr}(A \oplus B) = \mathrm{tr}\, A + \mathrm{tr}\, B$.
   (b) $\mathrm{tr}(A \otimes B) = (\mathrm{tr}\, A)(\mathrm{tr}\, B)$.
   (c) The spectrum of $A \oplus B$ is the union of the spectra of A and B.
   (d) The spectrum of $A \otimes B$ consists of all the scalars of the form $\alpha\beta$, with α and β in the spectrum of A and of B, respectively.

§ 56. Triangular form

It is now quite easy to prove the easiest one of the so-called canonical form theorems. Our assumption about the scalar field (namely, that it is algebraically closed) is still in force.

THEOREM 1. *If A is any linear transformation on an n-dimensional vector space $\mathcal{V}$, then there exist $n + 1$ subspaces $\mathfrak{M}_0, \mathfrak{M}_1, \cdots, \mathfrak{M}_{n-1}, \mathfrak{M}_n$ with the following properties:*

(i) *each $\mathfrak{M}_j$ $(j = 0, 1, \cdots, n - 1, n)$ is invariant under A,*

(ii) *the dimension of $\mathfrak{M}_j$ is j,*

(iii) *$(\mathbf{0} =) \mathfrak{M}_0 \subset \mathfrak{M}_1 \subset \cdots \subset \mathfrak{M}_{n-1} \subset \mathfrak{M}_n \ (= \mathcal{V})$.*

PROOF. If $n = 0$ or $n = 1$, the result is trivial; we proceed by induction, assuming that the statement is correct for $n - 1$. Consider the dual transformation A' on $\mathcal{V}'$; since it has at least one proper vector, say x', there exists a one-dimensional subspace $\mathfrak{M}$ invariant under it, namely, the set of all multiples of x'. Let us denote by $\mathfrak{M}_{n-1}$ the annihilator (in $\mathcal{V}'' = \mathcal{V}$) of $\mathfrak{M}$, $\mathfrak{M}_{n-1} = \mathfrak{M}^0$; then $\mathfrak{M}_{n-1}$ is an $(n - 1)$-dimensional subspace of $\mathcal{V}$, and $\mathfrak{M}_{n-1}$ is invariant under A. Consequently we may consider A as a linear transformation on $\mathfrak{M}_{n-1}$ alone, and we may find $\mathfrak{M}_0, \mathfrak{M}_1, \cdots, \mathfrak{M}_{n-2}$, $\mathfrak{M}_{n-1}$, satisfying the conditions (i), (ii), (iii). We write $\mathfrak{M}_n = \mathcal{V}$, and we are done.

The chief interest of this theorem comes from its matricial interpretation. Since $\mathfrak{M}_1$ is one-dimensional, we may find in it a vector $x_1 \neq 0$. Since $\mathfrak{M}_1 \subset \mathfrak{M}_2$, it follows that x_1 is also in $\mathfrak{M}_2$, and since $\mathfrak{M}_2$ is two-dimensional, we may find in it a vector x_2 such that x_1 and x_2 span $\mathfrak{M}_2$. We proceed in this way by induction, choosing vectors x_j so that $x_1, \cdots, x_j$ lie in $\mathfrak{M}_j$ and span $\mathfrak{M}_j$ for $j = 1, \cdots, n$. We obtain finally a basis $\mathfrak{X} = \{x_1, \cdots, x_n\}$ in $\mathcal{V}$; let us compute the matrix of A in this coordinate system. Since x_j is in $\mathfrak{M}_j$ and since $\mathfrak{M}_j$ is invariant under A, it follows that Ax_j must be a linear combination of $x_1, \cdots, x_j$. Hence in the expression

$$Ax_j = \sum_i \alpha_{ij} x_i$$

the coefficient of x_i must vanish whenever $i > j$; in other words, $i > j$ implies $\alpha_{ij} = 0$. Hence the matrix of A has the *triangular form*

$$[A] = \begin{bmatrix} \alpha_{11} & \alpha_{12} & \alpha_{13} & \cdots & \alpha_{1n} \\ 0 & \alpha_{22} & \alpha_{23} & \cdots & \alpha_{2n} \\ \cdot & \cdot & \cdot & \cdots & \cdot \\ 0 & 0 & 0 & \cdots & \alpha_{n-1,n} \\ 0 & 0 & 0 & \cdots & \alpha_{nn} \end{bmatrix}.$$

It is clear from this representation that $\det (A - \alpha_{ii}) = 0$ for $i = 1, \cdots$, n, so that the α_{ii} are the proper values of A, appearing on the main diagonal of $[A]$ with the proper multiplicities. We sum up as follows.

THEOREM 2. *If A is a linear transformation on an n-dimensional vector space $\mathcal{V}$, then there exists a basis $\mathfrak{X}$ in $\mathcal{V}$ such that the matrix $[A; \mathfrak{X}]$ is triangular; or, equivalently, if $[A]$ is any matrix, there exists a non-singular matrix $[B]$ such that $[B]^{-1}[A][B]$ is triangular.*

The triangular form is useful for proving many results about linear transformations. It follows from it, for example, that for any polynomial p, the proper values of $p(A)$, including their algebraic multiplicities, are precisely the numbers $p(\lambda)$, where λ runs through the proper values of A.

A large part of the theory of linear transformations is devoted to improving the triangularization result just obtained. The best thing a matrix can be is not triangular but *diagonal* (that is, $\alpha_{ij} = 0$ unless $i = j$); if a linear transformation is such that its matrix with respect to a suitable coordinate system is diagonal we shall call the transformation *diagonable*.

EXERCISES

1. Interpret the following matrices as linear transformations on $\mathbb{C}^2$ and, in each case, find a basis of $\mathbb{C}^2$ such that the matrix of the transformation with respect to that basis is triangular.

a) $\begin{pmatrix} 1 & 1 \\ 0 & 1 \end{pmatrix}$.

(b) $\begin{pmatrix} 1 & 1 \\ 1 & 0 \end{pmatrix}$.

(c) $\begin{pmatrix} 1 & 0 \\ 1 & 1 \end{pmatrix}$.

(d) $\begin{pmatrix} 1 & 1 \\ 1 & 1 \end{pmatrix}$.

(e) $\begin{pmatrix} 0 & 1 & 0 \\ 0 & 0 & 1 \\ 1 & 0 & 0 \end{pmatrix}$.

(f) $\begin{pmatrix} 0 & 1 & 1 \\ 0 & 0 & 1 \\ 1 & 0 & 0 \end{pmatrix}$.

2. Two commutative linear transformations on a finite-dimensional vector space $\mathcal{V}$ over an algebraically closed field can be simultaneously triangularized. In other words, if $AB = BA$, then there exists a basis $\mathcal{X}$ such that both $[A; \mathcal{X}]$ and $[B; \mathcal{X}]$ are triangular. [Hint: to imitate the proof in § 56, it is desirable to find a subspace $\mathfrak{M}$ of $\mathcal{V}$ invariant under both A and B. With this in mind, consider any proper value λ of A and examine the set of all solutions of $Ax = \lambda x$ for the role of $\mathfrak{M}$.]

3. Formulate and prove the analogues of the results of § 56 for triangular matrices below the diagonal (instead of above it).

4. Suppose that A is a linear transformation over an n-dimensional vector space. For every alternating n-linear form w, write $\bar{A}w$ for the function defined by

$$(\bar{A}w)(x_1, \cdots, x_n) = w(Ax_1, x_2, \cdots, x_n)$$
$$+ w(x_1, Ax_2, \cdots, x_n) + \cdots + w(x_1, x_2, \cdots, Ax_n).$$

Since $\bar{A}w$ is an alternating n-linear form, and, in fact, $\bar{A}$ is a linear transformation on the (one-dimensional) space of such forms, it follows that $\bar{A}w = \tau(A) \cdot w$, where $\tau(A)$ is a scalar.
   (a) $\tau(0) = 0$.
   (b) $\tau(1) = n$.
   (c) $\tau(A + B) = \tau(A) + \tau(B)$.
   (d) $\tau(\alpha A) = \alpha\tau(A)$.

(e) If the scalar field has characteristic zero and if A is a projection, then $\tau(A) = \rho(A)$.

(f) If (α_{ij}) is the matrix of A in some coordinate system, then $\tau(A) = \sum_i \alpha_{ii}$.

(g) $\tau(A') = \tau(A)$.

(h) $\tau(AB) = \tau(BA)$.

(i) For which permutations π of the integers $1, \cdots, k$ is it true that $\tau(A_1 \cdots A_k) = \tau(A_{\pi(1)} \cdots A_{\pi(k)})$ for all k-tuples $(A_1, \cdots, A_k)$ of linear transformations?

(j) If the field of scalars is algebraically closed, then $\tau(A) = \operatorname{tr} A$. (For this reason trace is usually defined to be τ; the most popular procedure is to use (f) as the definition.)

5. (a) Suppose that the scalar field has characteristic zero. Prove that if $E_1, \cdots, E_k$ and $E_1 + \cdots + E_k$ are projections, then $E_i E_j = 0$ whenever $i \neq j$. (Hint: from the fact that $\operatorname{tr}(E_1 + \cdots + E_k) = \operatorname{tr}(E_1) + \cdots + \operatorname{tr}(E_k)$ conclude that the range of $E_1 + \cdots + E_k$ is the direct sum of the ranges of $E_1, \cdots, E_k$.)

(b) If $A_1, \cdots, A_k$ are linear transformations on an n-dimensional vector space, and if $A_1 + \cdots + A_k = 1$ and $\rho(A_1) + \cdots + \rho(A_k) \leq n$, then each A_i is a projection and $A_i A_j = 0$ whenever $i \neq j$. (Start with $k = 2$ and proceed by induction; use a direct sum argument as in (a).)

6. (a) If A is a linear transformation on a finite-dimensional vector space over a field of characteristic zero, and if $\operatorname{tr} A = 0$, then there exists a basis $\mathfrak{X}$ such that if $[A; \mathfrak{X}] = (\alpha_{ij})$, then $\alpha_{ii} = 0$ for all i. (Hint: using the fact that A is not a scalar, prove first that there exists a vector x such that x and Ax are linearly independent. This proves that α_{11} can be made to vanish; proceed by induction.)

(b) Show that if the characteristic is not zero, the conclusion of (a) s false. (Hint: if the characteristic is 2, compute $BC - CB$, where $B = \begin{pmatrix} 0 & 1 \\ 0 & 0 \end{pmatrix}$ and $C = \begin{pmatrix} 0 & 0 \\ 1 & 0 \end{pmatrix}$.)

§ 57. Nilpotence

As an aid to getting a representation theorem more informative than the triangular one, we proceed to introduce and to study a very special but useful class of transformations. A linear transformation A is called *nilpotent* if there exists a strictly positive integer q such that $A^q = 0$; the least such integer q is the *index* of nilpotence.

THEOREM 1. *If A is a nilpotent linear transformation of index q on a finite-dimensional vector space $\mathcal{V}$, and if x_0 is a vector for which $A^{q-1}x_0 \neq 0$, then the vectors $x_0, Ax_0, \cdots, A^{q-1}x_0$ are linearly independent. If $\mathfrak{X}$ is the subspace spanned by these vectors, then there exists a subspace $\mathfrak{K}$ such that $\mathcal{V} = \mathfrak{X} \oplus \mathfrak{K}$ and such that the pair $(\mathfrak{X}, \mathfrak{K})$ reduces A.*

PROOF. To prove the asserted linear independence, suppose that $\sum_{i=0}^{q-1} \alpha_i A^i x_0 = 0$, and let j be the least index such that $\alpha_j \neq 0$. (We do not exclude the possibility $j = 0$.) Dividing through by $-\alpha_j$ and changing the notation in an obvious way, we obtain a relation of the form

$$A^j x_0 = \sum_{i=j+1}^{q-1} \alpha_i A^i x_0 = A^{j+1}\left(\sum_{i=j+1}^{q-1} \alpha_i A^{i-j-1} x_0\right) = A^{j+1}y.$$

It follows from the definition of q that

$$A^{q-1}x_0 = A^{q-j-1}A^jx_0 = A^{q-j-1}A^{j+1}y = A^qy = 0;$$

since this contradicts the choice of x_0, we must have $\alpha_j = 0$ for each j.

It is clear that $\mathcal{K}$ is invariant under A; to construct $\mathcal{K}$ we go by induction on the index q of nilpotence. If $q = 1$, the result is trivial; we now assume the theorem for $q - 1$. The range $\mathcal{R}$ of A is a subspace that is invariant under A; restricted to $\mathcal{R}$ the linear transformation A is nilpotent of index $q - 1$. We write $\mathcal{K}_0 = \mathcal{K} \cap \mathcal{R}$ and $y_0 = Ax_0$; then $\mathcal{K}_0$ is spanned by the linearly independent vectors $y_0, Ay_0, \cdots, A^{q-2}y_0$. The induction hypothesis may be applied, and we may conclude that $\mathcal{R}$ is the direct sum of $\mathcal{K}_0$ and some other invariant subspace $\mathcal{K}_0$.

We write $\mathcal{K}_1$ for the set of all vectors x such that Ax is in $\mathcal{K}_0$; it is clear that $\mathcal{K}_1$ is a subspace. The temptation is great to set $\mathcal{K} = \mathcal{K}_1$ and to attempt to prove that $\mathcal{K}$ has the desired properties. Unfortunately this need not be true; $\mathcal{K}$ and $\mathcal{K}_1$ need not be disjoint. (It is true, but we shall not use the fact, that the intersection of $\mathcal{K}$ and $\mathcal{K}_1$ is contained in the null-space of A.) That, in spite of this, $\mathcal{K}_1$ is useful is caused by the fact that $\mathcal{K} + \mathcal{K}_1 = \mathcal{V}$. To prove this, observe that Ax is in $\mathcal{R}$ for every x, and, consequently, $Ax = y + z$ with y in $\mathcal{K}_0$ and z in $\mathcal{K}_0$. The general element of $\mathcal{K}_0$ is a linear combination of $Ax_0, \cdots, A^{q-1}x_0$; hence we have

$$y = \sum_{i=1}^{q-1} \alpha_i A^i x_0 = A\left(\sum_{i=0}^{q-2} \alpha_{i+1} A^i x_0\right) = Ay_1,$$

where y_1 is in $\mathcal{K}$. It follows that $Ax = Ay_1 + z$, or $A(x - y_1) = z$, so that $A(x - y_1)$ is in $\mathcal{K}_0$. This means that $x - y_1$ is in $\mathcal{K}_1$, so that x is the sum of an element (namely y_1) of $\mathcal{K}$ and an element (namely $x - y_1$) of $\mathcal{K}_1$.

As far as disjointness is concerned, we can say at least that $\mathcal{K} \cap \mathcal{K}_0 = \Theta$. To prove this, suppose that x is in $\mathcal{K} \cap \mathcal{K}_0$, and observe first that Ax is in $\mathcal{K}_0$ (since x is in $\mathcal{K}$). Since $\mathcal{K}_0$ is also invariant under A, the vector Ax belongs to $\mathcal{K}_0$ along with x, so that $Ax = 0$. From this we infer that x is in $\mathcal{K}_0$. (Since x is in $\mathcal{K}$, we have $x = \sum_{i=0}^{q-1} \alpha_i A^i x_0$; and therefore $0 = Ax = \sum_{i=1}^{q-1} \alpha_{i-1} A^i x_0$; from the linear independence of the $A^j x_0$ it follows that $\alpha_0 = \cdots = \alpha_{q-2} = 0$, so that $x = \alpha_{q-1} A^{q-1}x_0$.) We have proved that if x belongs to $\mathcal{K} \cap \mathcal{K}_0$, then it belongs also to $\mathcal{K}_0 \cap \mathcal{K}_0$, and hence that $x = 0$.

The situation now is this: $\mathcal{K}$ and $\mathcal{K}_1$ together span $\mathcal{V}$, and $\mathcal{K}_1$ contains the two disjoint subspaces $\mathcal{K}_0$ and $\mathcal{K} \cap \mathcal{K}_1$. If we let $\mathcal{K}'_0$ be any complement of $\mathcal{K}_0 \oplus (\mathcal{K} \cap \mathcal{K}_1)$ in $\mathcal{K}_1$, that is, if

$$\mathcal{K}'_0 \oplus \mathcal{K}_0 \oplus (\mathcal{K} \cap \mathcal{K}_1) = \mathcal{K}_1,$$

then we may write $\mathcal{K} = \mathcal{K}'_0 \oplus \mathcal{K}_0$; we assert that this $\mathcal{K}$ has the desired properties. In the first place, $\mathcal{K} \subset \mathcal{K}_1$ and $\mathcal{K}$ is disjoint from $\mathcal{K} \cap \mathcal{K}_1$; it follows that $\mathfrak{K} \cap \mathfrak{K} = \mathfrak{O}$. In the second place, $\mathfrak{K} \oplus \mathfrak{K}$ contains both $\mathfrak{K}$ and $\mathfrak{K}_1$, so that $\mathfrak{K} \oplus \mathfrak{K} = \mathfrak{V}$. Finally, $\mathfrak{K}$ is invariant under A, since the fact that $\mathfrak{K} \subset \mathfrak{K}_1$ implies that $A\mathfrak{K} \subset \mathfrak{K}_0 \subset \mathfrak{K}$. The proof of the theorem is complete.

Later we shall need the following remark. If $\mathfrak{x}_0$ is any other vector for which $A^{q-1}\mathfrak{x}_0 \neq 0$, if $\tilde{\mathfrak{K}}$ is the subspace spanned by the vectors $\mathfrak{x}_0$, $A\mathfrak{x}_0$, $\cdots$, $A^{q-1}\mathfrak{x}_0$, and if, finally, $\tilde{\mathfrak{K}}$ is any subspace that together with $\tilde{\mathfrak{K}}$ reduces A, then the behavior of A on $\tilde{\mathfrak{K}}$ and $\tilde{\mathfrak{K}}$ is the same as its behavior on $\mathfrak{K}$ and $\mathfrak{K}$ respectively. (In other words, in spite of the apparent non-uniqueness in the statement of Theorem 1, everything is in fact uniquely determined up to isomorphisms.) The truth of this remark follows from the fact that the index of nilpotence of A on $\mathfrak{K}$ (r, say) is the same as the index of nilpotence of A on $\tilde{\mathfrak{K}}$ ($\tilde{r}$, say). This fact, in turn, is proved as follows. Since $A^r\mathfrak{V} = A^r\mathfrak{K} + A^r\mathfrak{K}$ and also $A^r\mathfrak{V} = A^r\tilde{\mathfrak{K}} + A^r\tilde{\mathfrak{K}}$ (these results depend on the invariance of all the subspaces involved), it follows that the dimensions of the right sides of these equations may be equated, and hence that $(q - r) + 0 = (q - r) + (\tilde{r} - r)$.

Using Theorem 1 we can find a complete geometric characterization of nilpotent transformations.

Theorem 2. *If A is a nilpotent linear transformation of index q on a finite-dimensional vector space $\mathfrak{V}$, then there exist positive integers r, q_1, $\cdots$, q_r and vectors x_1, $\cdots$, x_r such that* (i) $q_1 \geqq \cdots \geqq q_r$, (ii) *the vectors*

$$x_1, Ax_1, \cdots, A^{q_1-1}x_1,$$
$$x_2, Ax_2, \cdots, A^{q_2-1}x_2,$$
$$\cdot \quad \cdot \quad \cdot \quad \cdot \quad \cdot \quad \cdot \quad \cdot \quad \cdot \quad \cdot$$
$$x_r, Ax_r, \cdots, A^{q_r-1}x_r$$

form a basis for $\mathfrak{V}$, and (iii) $A^{q_1}x_1 = A^{q_2}x_2 = \cdots = A^{q_r}x_r = 0$. *The integers r, q_1, $\cdots$, q_r form a complete set of isomorphism invariants of A. If, in other words, B is any other nilpotent linear transformation on a finite-dimensional vector space $\mathfrak{W}$, then a necessary and sufficient condition that there exist an isomorphism T between $\mathfrak{V}$ and $\mathfrak{W}$ such that $TAT^{-1} = B$ is that the integers r, q_1, $\cdots$, q_r attached to B be the same as the ones attached to A.*

proof. We write $q_1 = q$ and we choose x_1 to be any vector for which $A^{q_1-1}x_1 \neq 0$. The subspace spanned by x_1, Ax_1, $\cdots$, $A^{q_1-1}x_1$ is invariant under A, and, by Theorem 1, possesses an invariant complement, which, naturally, has strictly lower dimension than $\mathfrak{V}$. On this complementary subspace A is nilpotent of index q_2, say; we apply the same reduction procedure to this subspace (beginning with a vector x_2 for which $A^{q_2-1}x_2 \neq 0$).

We continue thus by induction till we exhaust the space. This proves the existential part of the theorem; the remaining part follows from the uniqueness (up to isomorphisms) of the decomposition given by Theorem 1.

With respect to the basis $\{A^i x_j\}$ described in Theorem 2, the matrix of A takes on a particularly simple form. Every matrix element not on the diagonal just below the main diagonal vanishes (that is, $\alpha_{ij} \neq 0$ implies $j = i - 1$), and the elements below the main diagonal begin (at top) with a string of 1's followed by a single 0, then go on with another string of 1's followed by a 0, and continue so on to the end, with the lengths of the strings of 1's monotonely decreasing (or, at any rate, non-increasing).

Observe that our standing assumption about the algebraic closure of the field of scalars was not used in this section.

1. Does there exist a nilpotent transformation of index 3 on a 2-dimensional space?

2. (a) Prove that a nilpotent linear transformation on a finite-dimensional vector space has trace zero.

(b) Prove that if A and B are linear transformations (on the same finite-dimensional vector space) and if $C = AB - BA$, then $1 - C$ is not nilpotent.

3. Prove that if A is a nilpotent linear transformation of index q on a finite-dimensional vector space, then

$$\nu(A^{k+1}) + \nu(A^{k-1}) \leq 2\nu(A^k)$$

for $k = 1, \cdots, q - 1$.

4. If A is a linear transformation (on a finite-dimensional vector space over an algebraically closed field), then there exist linear transformations B and C such that $A = B + C$, B is diagonable, C is nilpotent, and $BC = CB$; the transformations B and C are uniquely determined by these conditions.

§ 58. Jordan form

It is sound geometric intuition that makes most of us conjecture that, for linear transformations, being invertible and being in some sense zero are exactly opposite notions. Our disappointment in finding that the range and the null-space need not be disjoint is connected with this conjecture. The situation can be straightened out by relaxing the sense in which we interpret "being zero"; for most practical purposes a linear transformation some power of which is zero (that is, a nilpotent transformation) is as zeroish as we can expect it to be. Although we cannot say that a linear transformation is either invertible or "zero" even in the extended sense of zeroness, we can say how any transformation is made up of these two extreme kinds.

THEOREM 1. *Every linear transformation A on a finite-dimensional vector space $\mathcal{V}$ is the direct sum of a nilpotent transformation and an invertible transformation.*

PROOF. We consider the null-space of the k-th power of A; this is a subspace $\mathfrak{N}_k = \mathfrak{N}(A^k)$. Clearly $\mathfrak{N}_1 \subset \mathfrak{N}_2 \subset \cdots$. We assert first that if ever $\mathfrak{N}_k = \mathfrak{N}_{k+1}$, then $\mathfrak{N}_k = \mathfrak{N}_{k+j}$ for all positive integers j. Indeed, if $A^{k+j}x = 0$, then $A^{k+1}A^{j-1}x = 0$, whence (by the fact that $\mathfrak{N}_k = \mathfrak{N}_{k+1}$) it follows that $A^k A^{j-1}x = 0$, and therefore that $A^{k+j-1}x = 0$. In other words, $\mathfrak{N}_{k+j}$ is contained in (and therefore equal to) $\mathfrak{N}_{k+j-1}$; induction on j establishes our assertion.

Since $\mathcal{V}$ is finite-dimensional, the subspaces $\mathfrak{N}_k$ cannot continue to increase indefinitely; let q be the smallest positive integer for which $\mathfrak{N}_q = \mathfrak{N}_{q+1}$. It is clear that $\mathfrak{N}_q$ is invariant under A (in fact each $\mathfrak{N}_k$ is such). We write $\mathfrak{R}_k = \mathfrak{R}(A^k)$ for the range of A^k (so that, again, it is clear that $\mathfrak{R}_q$ is invariant under A); we shall prove that $\mathcal{V} = \mathfrak{N}_q \oplus \mathfrak{R}_q$ and that A on $\mathfrak{N}_q$ is nilpotent, whereas on $\mathfrak{R}_q$ it is invertible.

If x is a vector common to $\mathfrak{N}_q$ and $\mathfrak{R}_q$, then $A^q x = 0$ and $x = A^q y$ for some y. It follows that $A^{2q}y = 0$, and hence, from the definition of q, that $x = A^q y = 0$. We have shown thus that the range and the null-space of A^q are disjoint; a dimensionality argument (see § 50, Theorem 1) shows that they span $\mathcal{V}$, so that $\mathcal{V}$ is their direct sum. It follows from the definitions of q and $\mathfrak{N}_q$ that A on $\mathfrak{N}_q$ is nilpotent of index q. If, finally, x is in $\mathfrak{R}_q$ (so that $x = A^q y$ for some y) and if $Ax = 0$, then $A^{q+1}y = 0$, whence $x = A^q y = 0$; this shows that A is invertible on $\mathfrak{R}_q$. The proof of Theorem 1 is complete.

The decomposition of A into its nilpotent and invertible parts is unique. Suppose, indeed, that $\mathcal{V} = \mathcal{K} \oplus \mathcal{K}$ so that A on $\mathcal{K}$ is nilpotent and A on $\mathcal{K}$ is invertible. Since $\mathcal{K} \subset \mathfrak{N}(A^k)$ for some k, it follows that $\mathcal{K} \subset \mathfrak{N}_q$, and, since $\mathcal{K} \subset \mathfrak{R}(A^k)$ for all k, it follows that $\mathcal{K} \subset \mathfrak{R}_q$; these facts together imply that $\mathcal{K} = \mathfrak{N}_q$ and $\mathcal{K} = \mathfrak{R}_q$.

We can now use our results on nilpotent transformations to study the structure of arbitrary transformations. The method of getting a nilpotent transformation out of an arbitrary one may seem like a conjuring trick, but it is a useful trick, which is often employed. What is essential is the guaranteed existence of proper values; for that reason we continue to assume that the scalar field is algebraically closed (see § 55).

THEOREM 2. *If A is a linear transformation on a finite-dimensional vector space $\mathcal{V}$, and if $\lambda_1, \cdots, \lambda_p$ are the distinct proper values of A with respective algebraic multiplicities $m_1, \cdots, m_p$, then $\mathcal{V}$ is the direct sum of p subspaces $\mathfrak{M}_1, \cdots, \mathfrak{M}_p$ of respective dimensions $m_1, \cdots, m_p$, such that each $\mathfrak{M}_j$ is invariant under A and such that $A - \lambda_j$ is nilpotent on $\mathfrak{M}_j$.*

PROOF. Take any fixed $j = 1, \cdots, p$, and consider the linear transformation $A_j = A - \lambda_j$. To A_j we may apply the decomposition of Theorem 1 to obtain subspaces $\mathfrak{M}_j$ and $\mathfrak{N}_j$ such that A_j is nilpotent on $\mathfrak{M}_j$ and invertible on $\mathfrak{N}_j$. Since $\mathfrak{M}_j$ is invariant under A_j, it is also invariant under $A_j + \lambda_j = A$. Hence, for every λ, the determinant of $A - \lambda$ is the product of the two corresponding determinants for the two linear transformations that A becomes when we consider it on $\mathfrak{M}_j$ and $\mathfrak{N}_j$ separately. Since the only proper value of A on $\mathfrak{M}_j$ is λ_j, and since A on $\mathfrak{N}_j$ does not have the proper value λ_j (that is, $A - \lambda_j$ is invertible on $\mathfrak{N}_j$), it follows that the dimension of $\mathfrak{M}_j$ is exactly m_j and that each of the subspaces $\mathfrak{M}_j$ is disjoint from the span of all the others. A dimension argument proves that $\mathfrak{M}_1 \oplus \cdots \oplus \mathfrak{M}_p = \mathcal{V}$ and thereby concludes the proof of the theorem.

We proceed to describe the principal results of this section and the preceding one in matricial language. If A is a linear transformation on a finite-dimensional vector space $\mathcal{V}$, then with respect to a suitable basis of $\mathcal{V}$, the matrix of A has the following form. Every element not on or immediately below the main diagonal vanishes. On the main diagonal there appear the distinct proper values of A, each a number of times equal to its algebraic multiplicity. Below any particular proper value there appear only 1's and 0's, and these in the following way: there are chains of 1's followed by a single 0, with the lengths of the chains decreasing as we read from top to bottom. This matrix is the *Jordan form* or the *classical canonical form* of A; we have $B = TAT^{-1}$ if and only if the classical canonical forms of A and B are the same except for the order of the proper values. (Thus, in particular, a linear transformation A is diagonable if and only if its classical canonical form is already diagonal, that is, if every chain of 1's has length zero.)

Let us introduce some notation. Let A have p distinct proper values $\lambda_1, \cdots, \lambda_p$, with algebraic multiplicities $m_1, \cdots, m_p$, as before; let the number of chains of 1's under λ_j be r_j, and let the lengths of these chains be $q_{j,1} - 1, q_{j,2} - 1, \cdots, q_{j,r_j} - 1$. The polynomial e_{ji} defined by $e_{ji}(\lambda) = (\lambda - \lambda_j)^{q_{j,i}}$ is called an *elementary divisor* of A of *multiplicity* $q_{j,i}$ belonging to the proper value λ_j. An elementary divisor is called *simple* if its multiplicity is 1 (so that the corresponding chain length is 0); we see that a linear transformation is diagonable if and only if its elementary divisors are simple.

To illustrate the power of Theorem 2 we make one application. We may express the fact that the transformation $A - \lambda_j$ on $\mathfrak{M}_j$ is nilpotent of index q_{j1} by saying that the transformation A on $\mathfrak{M}_j$ is annulled by the polynomial e_{j1}. It follows that A on $\mathcal{V}$ is annulled by the product of these polynomials (that is, by the product of the elementary divisors of the highest multiplicities); this product is called the *minimal polynomial* of A.

It is quite easy to see (since the index of nilpotence of $A - \lambda_j$ on $\mathfrak{M}_j$ is exactly $q_{j,1}$) that this polynomial is uniquely determined (up to a multiplicative factor) as the polynomial of smallest degree that annuls A. Since the characteristic polynomial of A is the product of all the elementary divisors, and therefore a multiple of the minimal polynomial, we obtain the *Hamilton-Cayley equation*: every linear transformation is annulled by its characteristic polynomial.

EXERCISES

1. Find the Jordan form of $\begin{pmatrix} 1 & 0 & 1 \\ 0 & 0 & 0 \\ 0 & 0 & -1 \end{pmatrix}$.

2. What is the maximum number of pairwise non-similar linear transformations on a three-dimensional vector space, each of which has the characteristic polynomial $(\lambda - 1)^3$?

3. Does every invertible linear transformation have a square root? (To say that A is a square root of B means, of course, that $A^2 = B$.)

4. (a) Prove that if ω is a cube root of 1 ($\omega \neq 1$), then the matrices

$$\begin{pmatrix} 0 & 1 & 0 \\ 0 & 0 & 1 \\ 1 & 0 & 0 \end{pmatrix} \quad \text{and} \quad \begin{pmatrix} 1 & 0 & 0 \\ 0 & \omega & 0 \\ 0 & 0 & \omega^2 \end{pmatrix}$$

are similar.
   (b) Discover and prove a generalization of (a) to higher dimensions.

5. (a) Prove that the matrices $\begin{pmatrix} 0 & 1 & \alpha \\ 0 & 0 & 1 \\ 0 & 0 & 0 \end{pmatrix}$ and $\begin{pmatrix} 0 & 1 & 0 \\ 0 & 0 & 1 \\ 0 & 0 & 0 \end{pmatrix}$ are similar.

   (b) Discover and prove a generalization of (a) to higher dimensions.

6. (a) Show that the matrices

$$\begin{pmatrix} 1 & 1 & 1 \\ 1 & 1 & 1 \\ 1 & 1 & 1 \end{pmatrix} \quad \text{and} \quad \begin{pmatrix} 3 & 0 & 0 \\ 0 & 0 & 0 \\ 0 & 0 & 0 \end{pmatrix}$$

are similar (over, say, the field of complex numbers).
   (b) Discover and prove a generalization of (a) to higher dimensions.

7. If two real matrices are similar over $\mathfrak{C}$, then they are similar over $\mathfrak{R}$.

8. Prove that every matrix is similar to its transpose.

9. If A and B are n-by-n matrices such that the $2n$-by-$2n$ matrices $\begin{pmatrix} A & 0 \\ 0 & A \end{pmatrix}$ and $\begin{pmatrix} B & 0 \\ 0 & B \end{pmatrix}$ are similar, then A and B are similar.

10. Which of the following matrices are diagonable (over the field of complex numbers)?

(a) $\begin{pmatrix} 0 & 0 & 1 \\ 1 & 0 & 0 \\ 0 & 1 & 0 \end{pmatrix}$,

(d) $\begin{pmatrix} 0 & 0 & 1 \\ 0 & 0 & 0 \\ 1 & 0 & 0 \end{pmatrix}$,

(b) $\begin{pmatrix} 0 & 0 & 1 \\ 0 & 0 & 0 \\ 0 & 0 & 0 \end{pmatrix}$,

(e) $\begin{pmatrix} 1 & 0 & 0 \\ 0 & 0 & 0 \\ 0 & 0 & 1 \end{pmatrix}$.

c) $\begin{pmatrix} 0 & 0 & 1 \\ 0 & 0 & 0 \\ -1 & 0 & 0 \end{pmatrix}$,

What about the field of real numbers?

11. Show that the matrix

$$\begin{bmatrix} 0 & 1 & 0 & 0 \\ 0 & 0 & 1 & 0 \\ 0 & 0 & 0 & 1 \\ 1 & 0 & 0 & 0 \end{bmatrix}$$

is diagonable over the field of complex numbers but not over the field of real numbers.

12. Let π be a permutation of the integers $\{1, \cdots, n\}$; if $x = (\xi_1, \cdots, \xi_n)$ is a vector in $\mathbb{C}^n$, write $Ax = (\xi_{\pi(1)}, \cdots, \xi_{\pi(n)})$. Prove that A is diagonable and find a basis with respect to which the matrix of A is diagonal.

13. Suppose that A is a linear transformation and that $\mathfrak{M}$ is a subspace invariant under A. Prove that if A is diagonable, then so also is the restriction of A to $\mathfrak{M}$.

14. Under what conditions on the complex numbers $\alpha_1, \cdots, \alpha_n$ is the matrix

$$\begin{bmatrix} 0 & \cdots & 0 & \alpha_1 \\ 0 & \cdots & \alpha_2 & 0 \\ \vdots & \cdots & \vdots & \vdots \\ \alpha_n & \cdots & 0 & 0 \end{bmatrix}$$

diagonable (over the field of complex numbers)?

15. Are the following assertions true or false?

(a) A real two-by-two matrix with a negative determinant is similar to a diagonal matrix.

(b) If A is a linear transformation on a complex vector space, and if $A^k = 1$ for some positive integer k, then A is diagonable.

(c) If A is a nilpotent linear transformation on a finite-dimensional vector space, then A is diagonable.

16. If A is a linear transformation on a finite-dimensional vector space over an algebraically closed field, and if every proper value of A has algebraic multiplicity 1, then A is diagonable.

17. If the minimal polynomial of a linear transformation A on an n-dimensional vector space has degree n, then A is diagonable.

18. Find the minimal polynomials of all projections and all involutions.

19. What is the minimal polynomial of the matrix

$$\begin{bmatrix} \lambda_1 & 0 & 0 & \cdots & 0 \\ 0 & \lambda_2 & 0 & \cdots & 0 \\ 0 & 0 & \lambda_3 & \cdots & 0 \\ \cdot & \cdot & \cdot & \cdot & \cdot \\ 0 & 0 & 0 & \cdots & \lambda_n \end{bmatrix} ?$$

20. (a) What is the minimal polynomial of the differentiation operator on $\mathcal{O}_n$?
(b) What is the minimal polynomial of the transformation A on $\mathcal{O}_n$ defined by $(Ax)(t) = x(t + 1)$?

21. If A is a linear transformation with minimal polynomial p, and if q is a polynomial such that $q(A) = 0$, then q is divisible by p.

22. (a) If A and B are linear transformations, if p is a polynomial such that $p(AB) = 0$, and if $q(t) = tp(t)$, then $q(BA) = 0$.
(b) What can be inferred from (a) about the relation between the minimal polynomials of AB and of BA?

23. A linear transformation is invertible if and only if the constant term of its minimal polynomial is different from zero.

CHAPTER III

ORTHOGONALITY

§ 59. Inner products

Let us now get our feet back on the ground. We started in Chapter I by pointing out that we wish to generalize certain elementary properties of certain elementary spaces such as $\mathfrak{R}^2$. In our study so far we have done this, but we have entirely omitted from consideration one aspect of $\mathfrak{R}^2$. We have studied the qualitative concept of linearity; what we have entirely ignored are the usual quantitative concepts of angle and length. In the present chapter we shall fill this gap; we shall superimpose on the vector spaces to be studied certain numerical functions, corresponding to the ordinary notions of angle and length, and we shall study the new structure (vector space plus given numerical function) so obtained. For the added depth of geometric insight we gain in this way, we must sacrifice some generality; throughout the rest of this book we shall have to assume that the underlying field of scalars is either the field $\mathfrak{R}$ of real numbers or the field $\mathfrak{C}$ of complex numbers.

For a clue as to how to proceed, we first inspect $\mathfrak{R}^2$. If $x = (\xi_1, \xi_2)$ and $y = (\eta_1, \eta_2)$ are any two points in $\mathfrak{R}^2$, the usual formula for the distance between x and y, or the length of the segment joining x and y, is $\sqrt{(\xi_1 - \eta_1)^2 + (\xi_2 - \eta_2)^2}$. It is convenient to introduce the notation

$$\| x \| = \sqrt{\xi_1{}^2 + \xi_2{}^2}$$

for the distance from x to the origin $0 = (0, 0)$; in this notation the distance between x and y becomes $\| x - y \|$.

So much, for the present, for lengths and distances; what about angles? It turns out that it is much more convenient to study, in the general case, not any of the usual measures of angles but rather their cosines. (Roughly speaking, the reason for this is that the angle, in the usual picture in the circle of radius one, is the length of a certain circular arc, whereas the cosine of the angle is the length of a line segment; the latter is much easier to relate to our preceding study of linear functions.) Suppose then that we let α be the angle between the segment from 0 to x and the positive ξ_1 axis, and let β be the angle between the segment from 0 to y and the same axis; the angle between the two vectors x and y is $\alpha - \beta$, so that its cosine is

$$\cos(\alpha - \beta) = \cos\alpha\cos\beta + \sin\alpha\sin\beta = \frac{\xi_1\eta_1 + \xi_2\eta_2}{\|x\| \cdot \|y\|}.$$

Consider the expression $\xi_1\eta_1 + \xi_2\eta_2$; by means of it we can express both angle and length by very simple formulas. We have already seen that if we know the distance between 0 and x for all x, then we can compute the distance between any x and y; we assert now that if for every pair of vectors x and y we are given the value of $\xi_1\eta_1 + \xi_2\eta_2$, then in terms of this value we may compute all distances and all angles. Indeed, if we take $x = y$, then $\xi_1\eta_1 + \xi_2\eta_2$ becomes $\xi_1{}^2 + \xi_2{}^2 = \|x\|^2$, and this takes care of lengths; the cosine formula above gives us the angle in terms of $\xi_1\eta_1 + \xi_2\eta_2$ and the two lengths $\|x\|$ and $\|y\|$. To have a concise notation, let us write, for $x = (\xi_1, \xi_2)$ and $y = (\eta_1, \eta_2)$,

$$\xi_1\eta_1 + \xi_2\eta_2 = (x, y);$$

what we said above is summarized by the relations

$$\text{distance from 0 to } x = \|x\| = \sqrt{(x, x)},$$
$$\text{distance from } x \text{ to } y = \|x - y\|,$$
$$\text{cosine of angle between } x \text{ and } y = \frac{(x, y)}{\|x\| \cdot \|y\|}.$$

The important properties of (x, y), considered as a numerical function of the pair of vectors x and y, are the following: it is symmetric in x and y, it depends linearly on each of its two variables, and (unless $x = 0$) the value of (x, x) is always strictly positive. (The notational conflict between the use of parentheses in (x, y) and in (ξ_1, ξ_2) is only apparent. It could arise in two-dimensional spaces only, and even there confusion is easily avoided.)

Observe for a moment the much more trivial picture in $\mathfrak{R}^1$. For $x = (\xi_1)$ and $y = (\eta_1)$ we should have, in this case, $(x, y) = \xi_1\eta_1$ (and it is for this reason that (x, y) is known as the *inner product* or *scalar product* of x and y). The angle between any two vectors is either 0 or π, so that its cosine is either $+1$ or -1. This shows up the much greater sensitivity of the function given by (x, y), which takes on all possible numerical values.

§ 60. Complex inner products

What happens if we want to consider $\mathbb{C}^2$ instead of $\mathbb{R}^2$? The generalization seems to lie right at hand; for $x = (\xi_1, \xi_2)$ and $y = (\eta_1, \eta_2)$ (where now the ξ's and η's may be complex numbers), we write $(x, y) = \xi_1\eta_1 + \xi_2\eta_2$, and we hope that the expressions $\| x \| = (x, x)$ and $\| x - y \|$ can be used as sensible measures of distance. Observe, however, the following strange phenomenon (where $i = \sqrt{-1}$):

$$\| ix \|^2 = (ix, ix) = i(x, ix) = i^2(x, x) = -\| x \|^2.$$

This means that if $\| x \|$ is positive, that is, if x is at a positive distance from the origin, then ix is not; in fact the distance from 0 to ix is imaginary. This is very unpleasant; surely it is reasonable to demand that whatever it is that is going to play the role of (x, y) in this case, it should have the property that for $x = y$ it never becomes negative. A formal remedy lies close at hand; we could try to write

$$(x, y) = \xi_1\bar{\eta}_1 + \xi_2\bar{\eta}_2$$

(where the bar denotes complex conjugation). In this definition the expression (x, y) loses much of its former beauty; it is no longer quite symmetric in x and y and it is no longer quite linear in each of its variables. But, and this is what prompted us to give our new definition,

$$(x, x) = \xi_1\bar{\xi}_1 + \xi_2\bar{\xi}_2 = |\xi_1|^2 + |\xi_2|^2$$

is surely never negative. It is a priori dubious whether a useful and elegant theory can be built up on the basis of a function that fails to possess so many of the properties that recommended it to our attention in the first place; the apparent inelegance will be justified in what follows by its success. A cheerful portent is this. Consider the space $\mathbb{C}^1$ (that is, the set of all complex numbers). It is impossible to draw a picture of any configuration in this space and then to be able to tell it apart from a configuration in $\mathbb{R}^2$, but conceptually it is clearly a different space. The analogue of (x, y) in this space, for $x = (\xi_1)$ and $y = (\eta_1)$, is given by $(x, y) = \xi_1\bar{\eta}_1$, and this expression does have a simple geometric interpretation. If we join x and y to the origin by straight line segments, (x, y) will not, to be sure, be the cosine of the angle between the two segments; it turns out that, for $\| x \| = \| y \| = 1$, its real part is exactly that cosine.

The complex conjugates that we were forced to introduce here will come back to plague us later; for the present we leave this heuristic introduction and turn to the formal work, after just one more comment on the notation. The similarity of the symbols (,) and [,], the one used here for inner product and the other used earlier for linear functionals, is not accidental. We shall show later that it is, in fact, only the presence of the complex conjugation in (,) that makes it necessary to use for it a symbol different from [,]. For the present, however, we cannot afford the luxury of confusing the two.

§ 61. Inner product spaces

DEFINITION. An *inner product* in a (real or complex) vector space is a (respectively, real or complex) numerically valued function of the ordered pair of vectors x and y, such that

(1) $$(x, y) = \overline{(y, x)},$$

(2) $$(\alpha_1 x_1 + \alpha_2 x_2, y) = \alpha_1(x_1, y) + \alpha_2(x_2, y),$$

(3) $$(x, x) \geqq 0; \quad (x, x) = 0 \text{ if and only if } x = 0.$$

An *inner product space* is a vector space with an inner product.

We observe that in the case of a real vector space, the conjugation in (1) may be ignored. In any case, however, real or complex, (1) implies that (x, x) is always real, so that the inequality in (3) makes sense. In an inner product space we shall use the notation

$$\sqrt{(x, x)} = \| x \|;$$

the number $\| x \|$ is called the *norm* or *length* of the vector x. A real inner product space is sometimes called a *Euclidean space*; its complex analogue is called a *unitary space*.

As examples of unitary spaces we may consider $\mathbb{C}^n$ and $\mathcal{O}$; in the first case we write, for $x = (\xi_1, \cdots, \xi_n)$ and $y = (\eta_1, \cdots, \eta_n)$,

$$(x, y) = \sum_{i=1}^{n} \xi_i \bar{\eta}_i;$$

and, in $\mathcal{O}$, we write

$$(x, y) = \int_0^1 x(t)\overline{y(t)} \, dt.$$

The modifications that convert these examples into Euclidean spaces (that is, real inner product spaces) are obvious.

In a unitary space we have

(2') $$(x, \alpha_1 y_1 + \alpha_2 y_2) = \bar{\alpha}_1(x, y_1) + \bar{\alpha}_2(x, y_2).$$

(To transform the left side of (2') into the right side, use (1), expand by (2), and use (1) again.) This fact, together with the definition of an inner product, explains the terminology sometimes used to describe properties

(1), (2), (3) (and their consequence (2'). According to that terminology (x, y) is a *Hermitian symmetric* (1), *conjugate bilinear* ((2) and (2')), and *positive definite* (3) form. In a Euclidean space the conjugation in (2') may be ignored along with the conjugation in (1); in that case (x, y) is called a symmetric, bilinear, and positive definite form. We observe that in either case, the conditions on (x, y) imply for $\| x \|$ the homogeneity property

$$\| \alpha x \| = | \alpha | \cdot \| x \|.$$

(Proof: $\| \alpha x \|^2 = (\alpha x, \alpha x) = \alpha \bar{\alpha} (x, x)$.)

§ 62. Orthogonality

The most important relation among the vectors of an inner product space is orthogonality. By definition, the vectors x and y are called *orthogonal* if $(x, y) = 0$. We observe that this relation is symmetric; since $(x, y) = \overline{(y, x)}$, it follows that (x, y) and (y, x) vanish together. If we recall the motivation for the introduction of (x, y), the terminology explains itself; two vectors are orthogonal (or perpendicular) if the angle between them is 90°, that is, if the cosine of the angle between them is 0. Two subspaces are called orthogonal if every vector in each is orthogonal to every vector in the other.

A set $\mathfrak{X}$ of vectors is *orthonormal* if whenever both x and y are in $\mathfrak{X}$ it follows that $(x, y) = 0$ or $(x, y) = 1$ according as $x \neq y$ or $x = y$. (If $\mathfrak{X}$ is finite, say $\mathfrak{X} = \{x_1, \cdots, x_n\}$, we have $(x_i, x_j) = \delta_{ij}$.) We call an orthonormal set *complete* if it is not contained in any larger orthonormal set.

To make our last definition in this connection, we observe first that an orthonormal set is linearly independent. Indeed, if $\{x_1, \cdots, x_k\}$ is any finite subset of an orthonormal set $\mathfrak{X}$, then $\sum_i \alpha_i x_i = 0$ implies that

$$0 = (\sum_i \alpha_i x_i, x_j) = \sum_i \alpha_i (x_i, x_j) = \sum_i \alpha_i \delta_{ij} = \alpha_j;$$

in other words, a linear combination of the x's can vanish only if all the coefficients vanish. From this we conclude that in a finite-dimensional inner product space the number of vectors in an orthonormal set is always finite, and, in fact, not greater than the linear dimension of the space. We define, in this case, the *orthogonal dimension* of the space, as the largest number of vectors an orthonormal set can contain.

Warning: for all we know at this stage, the concepts of orthogonality and orthonormal sets are vacuous. Trivial examples can be used to show that things are not so bad as all that; the vector 0, for instance, is always orthogonal to every vector, and, if the space contains a non-zero vector x, then the set consisting of $\dfrac{x}{\| x \|}$ alone is an orthonormal set. We grant that these examples are not very inspiring. For the present, however, we remain content with them; soon we shall see that there are always "enough" orthogonal vectors to operate with in comfort.

Observe also that we have no right to assume that the number of elements in a complete orthonormal set is equal to the orthogonal dimension. The point is this: if we had an orthonormal set with that many elements, it would clearly be complete; it is conceivable, just the same, that some other set contains fewer elements, but is still complete because its nasty structure precludes the possibility of extending it. These difficulties are purely verbal and will evaporate the moment we start proving things; they occur only because from among the several possibilities for the definition of completeness we had to choose a definite one, and we must prove its equivalence with the others.

We need some notation. If $\mathcal{E}$ is any set of vectors in an inner product space $\mathcal{V}$, we denote by $\mathcal{E}^{\perp}$ the set of all vectors in $\mathcal{V}$ that are orthogonal to every vector in $\mathcal{E}$. It is clear that $\mathcal{E}^{\perp}$ is a subspace of $\mathcal{V}$ (whether or not $\mathcal{E}$ is one), and that $\mathcal{E}$ is contained in $\mathcal{E}^{\perp\perp} = (\mathcal{E}^{\perp})^{\perp}$. It follows that the subspace spanned by $\mathcal{E}$ is contained in $\mathcal{E}^{\perp\perp}$. In case $\mathcal{E}$ is a subspace, we shall call $\mathcal{E}^{\perp}$ the *orthogonal complement* of $\mathcal{E}$. We use the sign in order to be reminded of orthogonality (or perpendicularity). In informal discussions, $\mathcal{E}^{\perp}$ might be pronounced as "E perp."

EXERCISES

1. Given four complex numbers α, β, γ, and δ, try to define an inner product in $\mathcal{C}^2$ by writing

$$(x, y) = \alpha\xi_1\bar{\eta}_1 + \beta\xi_2\bar{\eta}_1 + \gamma\xi_1\bar{\eta}_2 + \delta\xi_2\bar{\eta}_2$$

whenever $x = (\xi_1, \xi_2)$ and $y = (\eta_1, \eta_2)$. Under what conditions on α, β, γ, and δ does this equation define an inner product?

2. Prove that if x and y are vectors in a unitary space, then

$$4(x, y) = \| x + y \|^2 - \| x - y \|^2 + i \| x + iy \|^2 - i \| x - iy \|^2.$$

3. If inner product in $\mathcal{P}_{n+1}$ is defined by $(x, y) = \int_0^1 x(t)\overline{y(t)}\, dt$, and if $x_j(t) = t^j$, $= 0, \cdots, n - 1$, find a polynomial of degree n orthogonal to $x_0, x_1, \cdots, x_{n-1}$.

4. (a) Two vectors x and y in a real inner product space are orthogonal if and only if $\| x + y \|^2 = \| x \|^2 + \| y \|^2$.

(b) Show that (a) becomes false if "real" is changed to "complex."

(c) Two vectors x and y in a complex inner product space are orthogonal if and only if $\| \alpha x + \beta y \|^2 = \| \alpha x \|^2 + \| \beta y \|^2$ for all pairs of scalars α and β.

(d) If x and y are vectors in a real inner product space, and if $\| x \| = \| y \|$, then $x - y$ and $x + y$ are orthogonal. (Picture?) Discuss the corresponding statement for complex spaces.

(e) If x and y are vectors in an inner product space, then

$$\| x + y \|^2 + \| x - y \|^2 = 2\| x \|^2 + 2\| y \|^2.$$

Picture?

§ 63. Completeness

THEOREM 1. *If $\mathfrak{X} = \{x_1, \cdots, x_n\}$ is any finite orthonormal set in an inner product space, if x is any vector, and if $\alpha_i = (x, x_i)$, then (Bessel's inequality)*

$$\sum_i |\alpha_i|^2 \leq \| x \|^2.$$

The vector $x' = x - \sum_i \alpha_i x_i$ is orthogonal to each x_j and, consequently, to the subspace spanned by $\mathfrak{X}$.

PROOF. For the first assertion:

$$0 \leq \| x' \|^2 = (x', x') = (x - \sum_i \alpha_i x_i, \, x - \sum_j \alpha_j x_j)$$

$$= (x, x) - \sum_i \alpha_i(x_i, x) - \sum_j \bar{\alpha}_j(x, x_j) + \sum_i \sum_j \alpha_i \bar{\alpha}_j(x_i, x_j)$$

$$= \| x \|^2 - \sum_i |\alpha_i|^2 - \sum_i |\alpha_i|^2 + \sum_i |\alpha_i|^2$$

$$= \| x \|^2 - \sum_i |\alpha_i|^2;$$

for the second assertion:

$$(x', x_j) = (x, x_j) - \sum_i \alpha_i(x_i, x_j) = \alpha_j - \alpha_j = 0.$$

THEOREM 2. *If $\mathfrak{X}$ is any finite orthonormal set in an inner product space $\mathcal{V}$, the following six conditions on $\mathfrak{X}$ are equivalent to each other.*

(1) *The orthonormal set $\mathfrak{X}$ is complete.*

(2) *If $(x, x_i) = 0$ for $i = 1, \cdots, n$, then $x = 0$.*

(3) *The subspace spanned by $\mathfrak{X}$ is the whole space $\mathcal{V}$.*

(4) *If x is in $\mathcal{V}$, then $x = \sum_i (x, x_i)x_i$.*

(5) *If x and y are in $\mathcal{V}$, then (Parseval's identity)*

$$(x, y) = \sum_i (x, x_i)(x_i, y).$$

(6) *If x is in $\mathcal{V}$, then*

$$\| x \|^2 = \sum_i |(x, x_i)|^2.$$

PROOF. We shall establish the implications (1) $\Rightarrow$ (2) $\Rightarrow$ (3) $\Rightarrow$ (4) $\Rightarrow$ (5) $\Rightarrow$ (6) $\Rightarrow$ (1). Thus we first assume (1) and prove (2), then assume (2) to prove (3), and so on till we finally prove (1) assuming (6).

(1) $\Rightarrow$ (2). If $(x, x_i) = 0$ for all i and $x \neq 0$, then we may adjoin $x/\| x \|$ to $\mathfrak{X}$ and thus obtain an orthonormal set larger than $\mathfrak{X}$.

$(2) \Rightarrow (3)$. If there is an x that is not a linear combination of the x_i, then, by the second part of Theorem 1, $x' = x - \sum_i (x, x_i)x_i$ is different from 0 and is orthogonal to each x_i.

$(3) \Rightarrow (4)$. If every x has the form $x = \sum_j \alpha_j x_j$, then

$$(x, x_i) = \sum_j \alpha_j(x_j, x_i) = \alpha_i.$$

$(4) \Rightarrow (5)$. If $x = \sum_i \alpha_i x_i$ and $y = \sum_j \beta_j x_j$, with $\alpha_i = (x, x_i)$ and $\beta_j = (y, x_j)$, then

$$(x, y) = \left(\sum_i \alpha_i x_i, \sum_j \beta_j x_j\right) = \sum_i \alpha_i \bar{\beta}_i(x_i, x_j) = \sum_i \alpha_i \bar{\beta}_i.$$

$(5) \Rightarrow (6)$. Set $x = y$.

$(6) \Rightarrow (1)$. If $\mathfrak{X}$ were contained in a larger orthogonal set, say if x_0 is orthogonal to each x_i, then

$$\| x_0 \|^2 = \sum_i |(x_0, x_i)|^2 = 0,$$

so that $x_0 = 0$.

§ 64. Schwarz's inequality

THEOREM. *If x and y are vectors in an inner product space, then (Schwarz's inequality)*

$$|(x, y)| \leq \| x \| \cdot \| y \|.$$

PROOF. If $y = 0$, both sides vanish. If $y \neq 0$, then the set consisting of the vector $y/\| y \|$ is orthonormal, and, consequently, by Bessel's inequality

$$|(x, y/\| y \|)|^2 \leq \| x \|^2.$$

The Schwarz inequality has important arithmetic, geometric, and analytic consequences.

(1) In any inner product space we define the distance $\delta(x, y)$ between two vectors x and y by

$$\delta(x, y) = \| x - y \| = \sqrt{(x - y, x - y)}.$$

In order for δ to deserve to be called a distance, it should have the following three properties:

(i) $\delta(x, y) = \delta(y, x)$,

(ii) $\delta(x, y) \geq 0$; $\delta(x, y) = 0$ if and only if $x = y$,

(iii) $\delta(x, y) \leq \delta(x, z) + \delta(z, y)$.

(In a vector space it is also pleasant to be sure that distance is invariant under translations:

(iv) $\delta(x, y) = \delta(x + z, y + z)$.)

Properties (i), (ii), and (iv) are obviously possessed by the particular δ we defined; the only question is the validity of the "triangle inequality" (iii). To prove (iii), we observe that

$$\begin{aligned}
\| x + y \|^2 &= (x + y, x + y) = \| x \|^2 + (x, y) + (y, x) + \| y \|^2 \\
&= \| x \|^2 + (x, y) + \overline{(x, y)} + \| y \|^2 \\
&= \| x \|^2 + 2 \operatorname{Re}(x, y) + \| y \|^2 \\
&\leq \| x \|^2 + 2 |(x, y)| + \| y \|^2 \\
&\leq \| x \|^2 + 2\| x \| \cdot \| y \| + \| y \|^2 \\
&= (\| x \| + \| y \|)^2;
\end{aligned}$$

replacing x by $x - z$ and y by $z - y$, we obtain

$$\| x - y \| \leq \| x - z \| + \| z - y \|,$$

and this is equivalent to (iii). (We use $\operatorname{Re} \zeta$ to denote the real part of the complex number ζ; if $\zeta = \xi + i\eta$, with real ξ and η, then $\operatorname{Re} \zeta = \xi$. The imaginary part of ζ, that is, the real number η, is denoted by $\operatorname{Im} \zeta$.)

(2) In the Euclidean space $\mathfrak{R}^n$, the expression

$$\frac{(x, y)}{\| x \| \cdot \| y \|}$$

gives the cosine of the angle between x and y. The Schwarz inequality in this case merely amounts to the statement that the cosine of a real angle is ≤ 1.

(3) In the unitary space $\mathfrak{C}^n$, the Schwarz inequality becomes the so-called Cauchy inequality; it asserts that for any two sequences $(\xi_1, \cdots, \xi_n)$ and $(\eta_1, \cdots, \eta_n)$ of complex numbers, we have

$$\left| \sum_{i=1}^n \xi_i \bar{\eta}_i \right|^2 \leq \sum_{i=1}^n |\xi_i|^2 \cdot \sum_{i=1}^n |\eta_i|^2.$$

(4) In the space $\mathcal{P}$, the Schwarz inequality becomes

$$\left| \int_0^1 x(t)\overline{y(t)} \, dt \right|^2 \leq \int_0^1 |x(t)|^2 \, dt \cdot \int_0^1 |y(t)|^2 \, dt.$$

It is useful to observe that the relations mentioned in (1)–(4) above are not only analogous to the general Schwarz inequality, but actually consequences or special cases of it.

(5) We mention in passing that there is room between the two notions (general vector spaces and inner product spaces) for an intermediate concept of some interest. This concept is that of a *normed* vector space, a vector space in which there is an acceptable definition of length, but nothing is said about angles. A norm in a (real or complex) vector space is a numerically valued function $\| x \|$ of the vectors x such that $\| x \| \geq 0$ un-

less $x = 0$, $\| \alpha x \| = | \alpha | \cdot \| x \|$, and $\| x + y \| \leq \| x \| + \| y \|$. Our discussion so far shows that an inner product space is a normed vector space; the converse is not in general true. In other words, if all we are given is a norm satisfying the three conditions just given, it may not be possible to find an inner product for which (x, x) is identically equal to $\| x \|^2$. In somewhat vague but perhaps suggestive terms, we may say that the norm in an inner product space has an essentially "quadratic" character that norms in general need not possess.

§ 65. Complete orthonormal sets

THEOREM. *If $\mathcal{U}$ is an n-dimensional inner product space, then there exist complete orthonormal sets in $\mathcal{U}$, and every complete orthonormal set in $\mathcal{U}$ contains exactly n elements. The orthogonal dimension of $\mathcal{U}$ is the same as its linear dimension.*

PROOF. To people not fussy about hunting for an element in a possibly uncountable set, the existence of complete orthonormal sets is obvious. Indeed, we have already seen that orthonormal sets exist, so we choose one; if it is not complete, we may enlarge it, and if the resulting orthonormal set is still not complete, we enlarge it again, and we proceed in this way by induction. Since an orthonormal set may contain at most n elements, in at most n steps we shall reach a complete orthonormal set. This set spans the whole space (see § 63, Theorem 2, (1) $\Rightarrow$ (3)), and, since it is also linearly independent, it is a basis and therefore contains precisely n elements. This proves the first assertion of the theorem; the second assertion is now obvious from the definitions.

There is a constructive method of avoiding this crude induction, and since it sheds further light on the notions involved, we reproduce it here as an alternative proof of the theorem.

Let $\mathfrak{X} = \{x_1, \cdots, x_n\}$ be any basis in $\mathcal{U}$. We shall construct a complete orthonormal set $\mathcal{Y} = \{y_1, \cdots, y_n\}$ with the property that each y_j is a linear combination of $x_1, \cdots, x_j$. To begin the construction, we observe that $x_1 \neq 0$ (since $\mathfrak{X}$ is linearly independent) and we write $y_1 = x_1 / \| x_1 \|$. Suppose now that $y_1, \cdots, y_r$ have been found so that they form an orthonormal set and so that each y_j $(j = 1, \cdots, r)$ is a linear combination of $x_1, \cdots, x_j$. We write

$$z = x_{r+1} - (\alpha_1 y_1 + \cdots + \alpha_r y_r),$$

where the values of the scalars $\alpha_1, \cdots, \alpha_r$ are still to be determined. Since

$$(z, y_j) = (x_{r+1} - \sum_i \alpha_i y_i, y_j) = (x_{r+1}, y_j) - \alpha_j$$

for $j = 1, \cdots, r$, it follows that if we choose $\alpha_j = (x_{r+1}, y_j)$, then $(z, y_j) = 0$

for $j = 1, \cdots, r$. Since, moreover, z is a linear combination of x_{r+1} and $y_1, \cdots, y_r$, it is also a linear combination of x_{r+1} and $x_1, \cdots, x_r$. Finally z is different from zero, since $x_1, \cdots, x_r, x_{r+1}$ are linearly independent and the coefficient of x_{r+1} in the expression for z is not zero. We write $y_{r+1} = z/\| z \|$; clearly $\{y_1, \cdots, y_r, y_{r+1}\}$ is again an orthonormal set with all the desired properties, and the induction step is accomplished. We shall make use of the fact that not only is each y_j a linear combination of the x's with indices between 1 and j, but, vice versa, each x_j is a linear combination of the y's with indices between 1 and j. The method of converting a linear basis into a complete orthonormal set that we just described is known as the *Gram-Schmidt orthogonalization process*.

We shall find it convenient and natural, in inner product spaces, to work exclusively with such bases as are also complete orthonormal sets. We shall call such a basis an *orthonormal basis* or an *orthonormal coordinate system;* in the future, whenever we discuss bases that are not necessarily orthonormal, we shall emphasize this fact by calling them linear bases.

EXERCISES

1. Convert $\mathcal{P}_2$ into an inner product space by writing $(x, y) = \int_0^1 x(t)\overline{y(t)} \, dt$ whenever x and y are in $\mathcal{P}_2$, and find a complete orthonormal set in that space.

2. If x and y are orthogonal unit vectors (that is, $\{x, y\}$ is an orthonormal set), what is the distance between x and y?

3. Prove that if $|(x, y)| = \| x \| \cdot \| y \|$ (that is, if the Schwarz inequality reduces to an equality), then x and y are linearly dependent.

4. (a) Prove that the Schwarz inequality remains true if, in the definition of an inner product, "strictly positive" is replaced by "non-negative."
(b) Prove that for a "non-negative" inner product of the type mentioned in (a), the set of all those vectors x for which $(x, x) = 0$ is a subspace.
(c) Form the quotient space modulo the subspace mentioned in (b) and show that the given "inner product" induces on that quotient space, in a natural manner, an honest (strictly positive) inner product.
(d) Do the considerations in (a), (b), and (c) extend to normed spaces (with possibly no inner product)?

5. (a) Given a strictly positive number α, try to define a norm in $\mathcal{R}^2$ by writing

$$\| x \| = (|\xi_1|^\alpha + |\xi_2|^\alpha)^{1/\alpha}$$

whenever $x = (\xi_1, \xi_2)$. Under what conditions on α does this equation define a norm?

(b) Prove that the equation

$$\| x \| = \max \{|\xi_1|, |\xi_2|\}$$

defines a norm in $\mathcal{R}^2$.

(c) To which ones among the norms defined in (a) and (b) does there correspond an inner product in $\mathfrak{R}^2$ such that $\|x\|^2 = (x, x)$ for all x in $\mathfrak{R}^2$?

6. (a) Prove that a necessary and sufficient condition on a real normed space that there exist an inner product satisfying the equation $\|x\|^2 = (x, x)$ for all x is that

$$\|x + y\|^2 + \|x - y\|^2 = 2\|x\|^2 + 2\|y\|^2$$

for all x and y.

(b) Discuss the corresponding assertion for complex spaces.

(c) Prove that a necessary and sufficient condition on a norm in $\mathfrak{R}^2$ that there exist an inner product satisfying the equation $\|x\|^2 = (x, x)$ for all x in $\mathfrak{R}^2$ is that the locus of the equation $\|x\| = 1$ be an ellipse.

7. If $\{x_1, \cdots, x_n\}$ is a complete orthonormal set in an inner product space, and if $y_j = \sum_{i=1}^{j} x_i, j = 1, \cdots, n$, express in terms of the x's the vectors obtained by applying the Gram-Schmidt orthogonalization process to the y's.

§ 66. Projection theorem

Since a subspace of an inner product space may itself be considered as an inner product space, the theorem of the preceding section may be applied. The following result, called the *projection theorem*, is the most important application.

THEOREM. *If $\mathfrak{M}$ is any subspace of a finite-dimensional inner product space $\mathfrak{V}$, then $\mathfrak{V}$ is the direct sum of $\mathfrak{M}$ and $\mathfrak{M}^\perp$, and $\mathfrak{M}^{\perp\perp} = \mathfrak{M}$.*

PROOF. Let $\mathfrak{X} = \{x_1, \cdots, x_m\}$ be an orthonormal set that is complete in $\mathfrak{M}$, and let z be any vector in $\mathfrak{V}$. We write $x = \sum_i \alpha_i x_i$, where $\alpha_i = (z, x_i)$; it follows from § 63, Theorem 1, that $y = z - x$ is in $\mathfrak{M}^\perp$, so that z is the sum of two vectors, $z = x + y$, with x in $\mathfrak{M}$ and y in $\mathfrak{M}^\perp$. That $\mathfrak{M}$ and $\mathfrak{M}^\perp$ are disjoint is clear; if x belonged to both, then we should have $\|x\|^2 = (x, x) = 0$. It follows from the theorem of § 18 that $\mathfrak{V} = \mathfrak{M} \oplus \mathfrak{M}^\perp$.

We observe that in the decomposition $z = x + y$, we have

$$(z, x) = (x + y, x) = \|x\|^2 + (y, x) = \|x\|^2,$$

and, similarly,

$$(z, y) = \|y\|^2.$$

Hence, if z is in $\mathfrak{M}^{\perp\perp}$, so that $(z, y) = 0$, then $\|y\|^2 = 0$, so that $z \ (=x)$ is in $\mathfrak{M}$; in other words, $\mathfrak{M}^{\perp\perp}$ is contained in $\mathfrak{M}$. Since we already know that $\mathfrak{M}$ is contained in $\mathfrak{M}^{\perp\perp}$, the proof of the theorem is complete.

This kind of direct sum decomposition of an inner product space (via a subspace and its orthogonal complement) is of considerable geometric interest. We shall study the associated projections a little later; they turn out to be an interesting and important subclass of the class of all projections. At present we remark only on the connection with the Pythagorean theorem; since $(z, x) = \| x \|^2$ and $(z, y) = \| y \|^2$, we have

$$\| z \|^2 = (z, z) = (z, x) + (z, y) = \| x \|^2 + \| y \|^2.$$

In other words, the square of the hypotenuse is the sum of the squares of the sides. More generally, if $\mathfrak{M}_1, \cdots, \mathfrak{M}_k$ are pairwise orthogonal subspaces in an inner product space $\mathcal{V}$, and if $x = x_1 + \cdots + x_k$, with x_j in $\mathfrak{M}_j$ for $j = 1, \cdots, k$, then

$$\| x \|^2 = \| x_1 \|^2 + \cdots + \| x_k \|^2.$$

§ 67. Linear functionals

We are now in a position to study linear functionals on inner product spaces. For a general n-dimensional vector space the dual space is also n-dimensional and is therefore isomorphic to the original space. There is, however, no obvious natural isomorphism that we can set up; we have to wait for the second dual space to get back where we came from. The main point of the theorem we shall prove now is that in inner product spaces there is a "natural" correspondence between $\mathcal{V}$ and $\mathcal{V}'$; the only cloud on the horizon is that in general it is not quite an isomorphism.

THEOREM. *To any linear functional y' on a finite-dimensional inner product space $\mathcal{V}$ there corresponds a unique vector y in $\mathcal{V}$ such that $y'(x) = (x, y)$ for all x.*

PROOF. If $y' = 0$, we may choose $y = 0$; let us from now on assume that $y'(x)$ is not identically zero. Let $\mathfrak{M}$ be the subspace consisting of all vectors x for which $y'(x) = 0$, and let $\mathfrak{N} = \mathfrak{M}^\perp$ be the orthogonal complement of $\mathfrak{M}$. The subspace $\mathfrak{N}$ contains a non-zero vector y_0; multiplying by a suitable constant, we may assume that $\| y_0 \| = 1$. We write $y = \overline{y'(y_0)} \cdot y_0$. (The bar denotes complex conjugation, as usual; in case $\mathcal{V}$ is a real inner product space and not a unitary space, the bar may be omitted.) We do then have the desired relation

(1) $$y'(x) = (x, y)$$

at least for $x = y_0$ and for all x in $\mathfrak{M}$. For an arbitrary x in $\mathcal{V}$, we write $x_0 = x - \lambda y_0$, where

$$\lambda = \frac{y'(x)}{y'(y_0)} ;$$

then $y'(x_0) = 0$ and $x = x_0 + \lambda y_0$ is a linear combination of two vectors for each of which (1) is valid. From the linearity of both sides of (1) it follows that (1) holds for x, as was to be proved.

To prove uniqueness, suppose that $(x, y_1) = (x, y_2)$ for all x. It follows that $(x, y_1 - y_2) = 0$ for all x, and therefore in particular for $x = y_1 - y_2$, so that $\| y_1 - y_2 \|^2 = 0$ and $y_1 = y_2$.

The correspondence $y' \rightleftarrows y$ is a one-to-one correspondence between $\mathcal{U}$ and $\mathcal{U}'$, with the property that to $y'_1 + y'_2$ there corresponds $y_1 + y_2$, and to $\alpha y'$ there corresponds $\bar{\alpha} y$; for this reason we refer to it as a *conjugate isomorphism*. In spite of the fact that this conjugate isomorphism makes $\mathcal{U}'$ practically indistinguishable from $\mathcal{U}$, it is wise to keep the two conceptually separate. One reason for this is that we should like $\mathcal{U}'$ to be an inner product space along with $\mathcal{U}$; if, however, we follow the clue given by the conjugate isomorphism between $\mathcal{U}$ and $\mathcal{U}'$, the conjugation again causes trouble. Let y'_1 and y'_2 be any two elements of $\mathcal{U}'$; if $y'_1(x) = (x, y_1)$ and $y'_2(x) = (x, y_2)$, the temptation is great to write

$$(y'_1, y'_2) = (y_1, y_2).$$

A moment's reflection will show that this expression may not satisfy § 61, (2), and is therefore not a suitable inner product. The trouble arises in complex (that is, unitary) spaces only; we have, for example,

$$(\alpha y'_1, y'_2) = (\bar{\alpha} y_1, y_2) = \bar{\alpha}(y_1, y_2) = \bar{\alpha}(y'_1, y'_2).$$

The remedy is clear; we write

(2) $$\qquad\qquad (y'_1, y'_2) = \overline{(y_1, y_2)} = (y_2, y_1);$$

we leave it to the reader to verify that with this definition $\mathcal{U}'$ becomes an inner product space in all cases. We shall denote this inner product space by $\mathcal{U}^*$.

We remark that our troubles (if they can be called that) with complex conjugation have so far been more notational than conceptual; it is still true that the only difference between the theory of Euclidean spaces and the theory of unitary spaces is that an occasional bar appears in the latter. More profound differences between the two theories will arise when we go to study linear transformations.

§ 68. Parentheses versus brackets

It becomes necessary now to straighten out the relation between general vector spaces and inner product spaces. The theorem of the preceding section shows that, as long as we are careful about complex conjugation, (x, y) can completely take the place of $[x, y]$. It might seem that it would have been desirable to develop the entire subject of general vector spaces in such a way that the concept of orthogonality in a unitary space becomes not merely an analogue but a special case of some previously studied general relation between vectors and functionals. One way, for example, of avoiding the unpleasantness of conjugation (or, rather, of shifting it to a less conspicuous position) would have been to define the dual space of a complex vector space as the set of conjugate linear functionals, that is, the set of numerically valued functions y for which

$$y(\alpha_1 x_1 + \alpha_2 x_2) = \bar{\alpha}_1 y(x_1) + \bar{\alpha}_2 y(x_2).$$

Because it seemed pointless (and contrary to common usage) to introduce this complication into the general theory, we chose instead the roundabout way that we just traveled. Since from now on we shall deal with inner product spaces only, we ask the reader mentally to revise all the preceding work by replacing, throughout, the bracket $[x, y]$ by the parenthesis (x, y). Let us examine the effect of this change on the theorems and definitions of the first two chapters.

The replacement of $\mathcal{U}'$ by $\mathcal{U}^*$ is merely a change of notation; the new symbol is supposed to remind us that something new (namely, an inner product) has been added to $\mathcal{U}'$. Of a little more interest is the (conjugate) isomorphism between $\mathcal{U}$ and $\mathcal{U}^*$; by means of it the theorems of § 15, asserting the existence of linear functionals with various properties, may now be interpreted as asserting the existence of certain vectors in $\mathcal{U}$ itself. Thus, for example, the existence of a dual basis to any given basis $\mathcal{X} = \{x_1, \cdots, x_n\}$ implies now the existence of a basis $\mathcal{Y} = \{y_1, \cdots, y_n\}$ (of $\mathcal{U}$) with the property that $(x_i, y_j) = \delta_{ij}$.

More exciting still is the implied replacement of the annihilator $\mathfrak{M}^0$ of a subspace $\mathfrak{M}$ ($\mathfrak{M}^0$ lying in $\mathcal{U}'$ or $\mathcal{U}^*$) by the orthogonal complement $\mathfrak{M}^\perp$ (lying, along with $\mathfrak{M}$, in $\mathcal{U}$). The most radical new development, however, concerns the adjoint of a linear transformation. Thus we may write the analogue of § 44, (1), and corresponding to every linear transformation A on $\mathcal{U}$ we may define a linear transformation A^* by writing

$$(Ax, y) = (x, A^*y)$$

for every x. It follows from this definition that A^* is again a linear transformation defined on the same vector space $\mathcal{U}$, but, because of the Hermitian symmetry of (x, y), the relation between A and A^* is not quite the same as the relation between A and A'. The most notable difference is that (in a unitary space) $(\alpha A)^* = \bar{\alpha} A^*$ (and not $(\alpha A)^* = \alpha A^*$). Associated with this phenomenon is the fact that if the matrix of A, with respect to some fixed basis, is (α_{ij}), then the matrix of A^*, with respect to the dual basis, is not (α_{ji}) but $(\bar{\alpha}_{ji})$. For determinants we do not have det $A^* = $ det A but det $A^* = \overline{\det A}$, and, consequently, the proper values of A^* are not the same as those of A, but rather their conjugates. Here, however, the differences stop. All the other results of § 44 on the anti-isomorphic nature of the correspondence $A \rightleftarrows A^*$ are valid; the identity $A = A^{**}$ is strictly true and does not need the help of an isomorphism to interpret it.

Presently we shall discuss linear transformations on inner product spaces and we shall see that the principal new feature that differentiates their study from the discussion of Chapter II is the possibility of comparing A and A^* as linear transformations on the same space, and of investigating those classes of linear transformations that bear a particularly simple relation to their adjoints.

§ 69. Natural isomorphisms

There is now only one more possible doubt that the reader might (or, at any rate, should) have. Many of our preceding results were consequences of such reflexivity relations as $A^{**} = A$; do these remain valid after the brackets-to-parentheses revolution? More to the point is the following way of asking the question. Everything we say about a unitary space $\mathcal{V}$ must also be true about the unitary space $\mathcal{V}^*$; in particular it is also in a natural conjugate isomorphic relation with its dual space $\mathcal{V}^{**}$. If now to every vector in $\mathcal{V}$ we make correspond a vector in $\mathcal{V}^{**}$, by first applying the natural conjugate isomorphism from $\mathcal{V}$ to $\mathcal{V}^*$ and then going the same way from $\mathcal{V}^*$ to $\mathcal{V}^{**}$, then this mapping is a rival for the title of natural mapping from $\mathcal{V}$ to $\mathcal{V}^{**}$, a title already awarded in Chapter I to a seemingly different correspondence. What is the relation between the two natural correspondences? Our statements about the coincidence, except for trivial modifications, of the parenthesis and bracket theories, are really justified by the fact, which we shall n ow prove, that the two mappings are the same. (It should not be surprising, since $\bar{\bar{\alpha}} = \alpha$, that after two applications the bothersome conjugation disappears.) The proof is shorter than the introduction to it.

Let y_0 be any element of $\mathcal{V}$; to it there corresponds the linear functional y_0^* in $\mathcal{V}^*$, defined by $y_0^*(x) = (x, y_0)$, and to y_0^*, in turn, there corresponds the linear functional y_0^{**} in $\mathcal{V}^{**}$, defined by $y_0^{**}(y^*) = (y^*, y_0^*)$. Both these correspondences are given by the mapping introduced in this chapter. Earlier (see § 16) the correspondent y_0^{**} in $\mathcal{V}^{**}$ of y_0 in $\mathcal{V}$ was defined by $y_0^{**}(y^*) = y^*(y_0)$ for all y^* in $\mathcal{V}^*$; we must show that y_0^{**}, as we here defined it, satisfies this identity. Let y^* be any linear functional on $\mathcal{V}$ (that is, any element of $\mathcal{V}^*$); we have

$$y_0^{**}(y^*) = (y^*, y_0^*) = (y_0, y) = y^*(y_0).$$

(The middle equality comes from the definition of inner product in $\mathcal{V}^*$.) This settles all our problems.

1. If $\mathfrak{M}$ and $\mathfrak{N}$ are subspaces of a finite-dimensional inner product space, then

$$(\mathfrak{M} + \mathfrak{N})^{\perp} = \mathfrak{M}^{\perp} \cap \mathfrak{N}^{\perp}$$

and

$$(\mathfrak{M} \cap \mathfrak{N})^{\perp} = \mathfrak{M}^{\perp} + \mathfrak{N}^{\perp}.$$

2. If $y'(x) = \frac{1}{3}(\xi_1 + \xi_2 + \xi_3)$ for each $x = (\xi_1, \xi_2, \xi_3)$ in $\mathbb{C}^3$, find a vector y in $\mathbb{C}^3$ such that $y'(x) = (x, y)$.

3. If y is a vector in an inner product space, if A is a linear transformation on that space, and if $f(x) = \overline{(y, Ax)}$ for every vector x, then f is a linear functional; find a vector y^* such that $f(x) = (x, y^*)$ for every x.

4. (a) If A is a linear transformation on a finite-dimensional inner product space, then tr $(A^*A) \geqq 0$; a necessary and sufficient condition that tr $(A^*A) = 0$ is that $A = 0$. (Hint: look at matrices.) This property of traces can often be used to obtain otherwise elusive algebraic facts about products of transformations and their adjoints.

(b) Prove by a trace argument, and also directly, that if $A_1, \cdots, A_k$ are linear transformations on a finite-dimensional inner product space and if $\sum_{j=1}^{k} A_j^*A_j = 0$, then $A_1 = \cdots = A_k = 0$.

(c) If $A^*A = B^*B - BB^*$, then $A = 0$.

(d) If A^* commutes with A and if A commutes with B, then A^* commutes with B. (Hint: if $C = A^*B - BA^*$ and $D = AB - BA$, then tr $(C^*C) = $ tr (D^*D) $+ $ tr $[(A^*A - AA^*)(B^*B - BB^*)]$.)

5. (a) Suppose that $\mathfrak{IC}$ is a unitary space, and form the set of all ordered pairs $\langle x, y \rangle$ with x and y in $\mathfrak{IC}$ (that is, the direct sum of $\mathfrak{IC}$ with itself). Prove that the equation

$$(\langle x_1, y_1 \rangle, \langle x_2, y_2 \rangle) = (x_1, x_2) + (y_1, y_2)$$

defines an inner product in the direct sum $\mathfrak{IC} \oplus \mathfrak{IC}$.

(b) If U is defined by $U\langle x, y \rangle = \langle y, -x \rangle$, then $U^*U = 1$.

(c) The *graph* of a linear transformation A on $\mathfrak{IC}$ is the set of all those elements $\langle x, y \rangle$ of $\mathfrak{IC} \oplus \mathfrak{IC}$ for which $y = Ax$. Prove that the graph of every linear transformation on $\mathfrak{IC}$ is a subspace of $\mathfrak{IC} \oplus \mathfrak{IC}$.

(d) If A is a linear transformation on $\mathfrak{IC}$ with graph $\mathfrak{G}$, then the graph of A^* is the orthogonal complement (in $\mathfrak{IC} \oplus \mathfrak{IC}$) of the image under U (see (b)) of the graph of A.

6. (a) If for every linear transformation A on a finite-dimensional inner product space $N(A) = \sqrt{\text{tr} (A^*A)}$, then N is a norm (on the space of all linear transformations).

(b) Is the norm N induced by an inner product?

7. (a) Two linear transformations A and B on an inner product space are called *congruent* if there exists an invertible linear transformation P such that $B = P^*AP$. (The concept is frequently defined for the "quadratic forms" associated with linear transformations and not for the linear transformations themselves; this is largely a matter of taste. Note that if $\alpha(x) = (Ax, x)$ and $\beta(x) = (Bx, x)$, then $B = P^*AP$ implies that $\beta(x) = \alpha(Px)$.) Prove that congruence is an equivalence relation.

(b) If A and B are congruent, then so also are A^* and B^*.

(c) Does there exist a linear transformation A such that A is congruent to a scalar α, but $A \neq \alpha$?

(d) Do there exist linear transformations A and B such that A and B are congruent, but A^2 and B^2 are not?

(e) If two invertible transformations are congruent, then so are their inverses.

§ 70. Self-adjoint transformations

Let us now study the algebraic structure of the class of all linear transformations on an inner product space $\mathcal{V}$. In many fundamental respects this class resembles the class of all complex numbers. In both systems, notions of addition, multiplication, 0, and 1 are defined and have similar properties, and in both systems there is an involutory anti-automorphism of the system onto itself (namely, $A \rightarrow A^*$ and $\zeta \rightarrow \bar{\zeta}$). We shall use this analogy as a heuristic principle, and we shall attempt to carry over to linear transformations some well-known concepts from the complex domain. We shall be hindered in this work by two difficulties in the theory of linear transformations, of which, possibly surprisingly, the second is much more serious; they are the impossibility of unrestricted division and the non-commutativity of general linear transformations.

The three most important subsets of the complex number plane are the set of real numbers, the set of positive real numbers, and the set of numbers of absolute value one. We shall now proceed systematically to use our heuristic analogy of transformations with complex numbers, and to try to discover the analogues among transformations of these well-known numerical concepts.

When is a complex number real? Clearly a necessary and sufficient condition for the reality of ζ is the validity of the equation $\zeta = \bar{\zeta}$. We might accordingly (remembering that the analogue of the complex conjugate for linear transformations is the adjoint) define a linear transformation A to be real if $A = A^*$. More commonly linear transformations A for which $A = A^*$ are called *self-adjoint*; in real inner product spaces the usual word is *symmetric*, and, in complex inner product spaces, *Hermitian*. We shall see that self-adjoint transformations do indeed play the same role as real numbers.

It is quite easy to characterize the matrix of a self-adjoint transformation with respect to an orthonormal basis $\mathcal{X} = \{x_1, \cdots, x_n\}$. If the matrix of A is (α_{ij}), then we know that the matrix of A^* with respect to the dual basis of $\mathcal{X}$ is (α_{ij}^*), where $\alpha_{ij}^* = \overline{\alpha_{ji}}$; since an orthonormal basis is self-dual and since $A = A^*$, we have

$$\alpha_{ij} = \overline{\alpha_{ji}}.$$

We leave it to the reader to verify the converse: if we define a linear transformation A by means of a matrix (α_{ij}) and an arbitrary orthonormal coordinate system $\mathfrak{X} = \{x_1, \cdots, x_n\}$, via the usual equations

$$A(\textstyle\sum_j \xi_j x_j) = \sum_i \eta_i x_i,$$

$$\eta_i = \textstyle\sum_j \alpha_{ij}\xi_j,$$

and if the matrix (α_{ij}) is such that $\alpha_{ij} = \overline{\alpha_{ji}}$, then A is self-adjoint.

The algebraic rules for the manipulation of self-adjoint transformations are easy to remember if we think of such transformations as the analogues of real numbers. Thus, if A and B are self-adjoint, so is $A + B$; if A is self-adjoint and different from 0, and if α is a non-zero scalar, then a necessary and sufficient condition that αA be self-adjoint is that α be real; and if A is invertible, then both or neither of A and A^{-1} are self-adjoint. The place where something always goes wrong is in multiplication; the product of two self-adjoint transformations need not be self-adjoint. The positive facts about products are given by the following two theorems.

THEOREM 1. *If A and B are self-adjoint, then a necessary and sufficient condition that AB (or BA) be self-adjoint is that $AB = BA$ (that is that A and B commute).*

PROOF. If $AB = BA$, then $(AB)^* = B^*A^* = BA = AB$. If $(AB)^* = AB$, then $AB = (AB)^* = B^*A^* = BA$.

THEOREM 2. *If A is self-adjoint, then B^*AB is self-adjoint for all B; if B is invertible and B^*AB is self-adjoint, then A is self-adjoint.*

PROOF. If $A = A^*$, then $(B^*AB)^* = B^*A^*B^{**} = B^*AB$. If B is invertible and $B^*AB = (B^*AB)^* = B^*A^*B$, then (multiply by B^{*-1} on the left and B^{-1} on the right) $A = A^*$.

A complex number ζ is purely imaginary if and only if $\bar{\zeta} = -\zeta$. The corresponding concept for linear transformations is identified by the word *skew*; if a linear transformation A on an inner product space is such that $A^* = -A$, then A is called *skew symmetric* or *skew Hermitian* according as the space is real or complex. Here is some evidence for the thoroughgoing nature of our analogy between complex numbers and linear transformations: an arbitrary linear transformation A may be expressed, in one and only one way, in the form $A = B + C$, where B is self-adjoint and C is skew. (The representation of A in this form is sometimes called the *Cartesian decomposition* of A.) Indeed, if we write

(1) $$B = \frac{A + A^*}{2},$$

(2) $$C = \frac{A - A^*}{2},$$

then we have $B^* = \dfrac{A^* + A}{2} = B$ and $C^* = \dfrac{A^* - A}{2} = -C$, and, of course, $A = B + C$. From this proof of the existence of the Cartesian decomposition, its uniqueness is also clear; if we do have $A = B + C$, then $A^* = B - C$, and, consequently, A, B, and C are again connected by (1) and (2).

In the complex case there is a simple way of getting skew Hermitian transformations from Hermitian ones, and vice versa: just multiply by $i (= \sqrt{-1})$. It follows that, in the complex case, every linear transformation A has a unique representation in the form $A = B + iC$, where B and C are Hermitian. We shall refer to B and C as the real and imaginary parts of A.

EXERCISES

1. Give an example of two self-adjoint transformations whose product is not self-adjoint.

2. Consider the space $\mathcal{P}_n$ with the inner product given by $(x, y) = \int_0^1 x(t)\overline{y(t)}\, dt$.

(a) Is the multiplication operator T (defined by $(Tx)(t) = tx(t)$) self-adjoint?
(b) Is the differentiation operator D self-adjoint?

3. (a) Prove that the equation $(x, y) = \sum_{j=0}^n x\left(\dfrac{j}{n}\right) \overline{y\left(\dfrac{j}{n}\right)}$ defines an inner product in the space $\mathcal{P}_n$.

(b) Is the multiplication operator T (defined by $(Tx)(t) = tx(t)$) self-adjoint (with respect to the inner product defined in (a))?

(c) Is the differentiation operator D self-adjoint?

4. If A and B are linear transformations such that A and AB are self-adjoint and such that $\mathfrak{R}(A) \subset \mathfrak{R}(B)$, then there exists a self-adjoint transformation C such that $CA = B$.

5. If A and B are congruent and A is skew, does it follow that B is skew?

6. If A is skew, does it follow that so is A^2? How about A^3?

7. If both A and B are self-adjoint, or else if both are skew, then $AB + BA$ is self-adjoint and $AB - BA$ is skew. What happens if one of A and B is self-adjoint and the other skew?

8. If A is a skew-symmetric transformation on a Euclidean space, then $(Ax, x) = 0$ for every vector x. Converse?

9. If A is self-adjoint, or skew, and if $A^2 x = 0$, then $Ax = 0$.

10. (a) If A is a skew-symmetric transformation on a Euclidean space of odd dimension, then $\det A = 0$.

(b) If A is a skew-symmetric transformation on a finite-dimensional Euclidean space, then $\rho(A)$ is even.

§ 71. Polarization

Before continuing with the program of studying the analogies between complex numbers and linear transformations, we take time out to pick up some important auxiliary results about inner product spaces.

THEOREM 1. *A necessary and sufficient condition that a linear transformation A on an inner product space be 0 is that $(Ax, y) = 0$ for all x and y.*

PROOF. The necessity of the condition is obvious; sufficiency follows from setting y equal to Ax.

THEOREM 2. *A necessary and sufficient condition that a self-adjoint linear transformation A on an inner product space A be 0 is that $(Ax, x) = 0$ for all x.*

PROOF. Necessity is obvious. The proof of sufficiency begins by verifying the identity

$$(1) \quad (Ax, y) + (Ay, x) = (A(x + y), (x + y)) - (Ax, x) - (Ay, y).$$

(Expand the first term on the right side.) Since A is self-adjoint, the left side of this equation is equal to $2\,\mathrm{Re}\,(Ax, y)$. The assumed condition implies that the right side vanishes, and hence that $\mathrm{Re}\,(Ax, y) = 0$. At this point it is necessary to split the proof into two cases. If the inner product space is real (that is, A is symmetric), then (Ax, y) is real, and therefore $(Ax, y) = 0$. If the inner product space is complex (that is, A is Hermitian), then we find a complex number θ such that $|\theta| = 1$ and $\theta(Ax, y) = |(Ax, y)|$. (Here x and y are temporarily fixed.) The result we already have, applied to θx in place of x, yields $0 = \mathrm{Re}\,(A(\theta x), y) = \mathrm{Re}\,\theta(Ax, y) = \mathrm{Re}\,|(Ax, y)| = |(Ax, y)|$. In either case, therefore, $(Ax, y) = 0$ for all x and y, and the desired result follows from Theorem 1.

It is useful to ask how important is the self-adjointness of A in Theorem 2; the answer is that in the complex case it is not important at all.

THEOREM 3. *A necessary and sufficient condition that a linear transformation A on a unitary space be 0 is that $(Ax, x) = 0$ for all x.*

PROOF. As before, necessity is obvious. For the proof of sufficiency we use the so-called *polarization* identity:

$$(2) \quad \alpha\bar{\beta}(Ax, y) + \bar{\alpha}\beta(Ay, x)$$
$$= (A(\alpha x + \beta y), (\alpha x + \beta y)) - |\alpha|^2(Ax, x) - |\beta|^2(Ay, y).$$

(Just as for (1), the proof consists of expanding the first term on the right.) If (Ax, x) is identically zero, then we obtain, first choosing $\alpha = \beta = 1$, and then $\alpha = i \ (= \sqrt{-1}\,), \beta = 1$

$$(Ax, y) + (Ay, x) = 0$$

$$i(Ax, y) - i(Ay, x) = 0.$$

Dividing the second of these two equations by i and then forming their arithmetic mean, we see that $(Ax, y) = 0$ for all x and y, so that, by Theorem 1, $A = 0$.

This process of polarization is often used to get information about the "bilinear form" (Ax, y) when only knowledge of the "quadratic form" (Ax, x) is assumed.

It is important to observe that, despite its seeming innocence, Theorem 3 makes very essential use of the complex number system; it and many of its consequences fail to be true for real inner product spaces. The proof, of course, breaks down at our choice of $\alpha = \sqrt{-1}$. For an example consider a 90° rotation of the plane; it clearly has the property that it sends every vector x into a vector orthogonal to x.

We have seen that Hermitian transformations play the same role as real numbers; the following theorem indicates that they are tied up with the concept of reality in deeper ways than through the formal analogy that suggested their definition.

THEOREM 4. *A necessary and sufficient condition that a linear transformation A on a unitary space be Hermitian is that (Ax, x) be real for all x.*

PROOF. If $A = A^*$, then

$$(Ax, x) = (x, A^*x) = (x, Ax) = \overline{(Ax, x)},$$

so that (Ax, x) is equal to its own conjugate and is therefore real. If, conversely, (Ax, x) is always real, then

$$(Ax, x) = \overline{(Ax, x)} = (x, A^*x) = (A^*x, x),$$

so that $([A - A^*]x, x) = 0$ for all x, and, by Theorem 3, $A = A^*$.

Theorem 4 is false for real inner product spaces. This is to be expected, for, in the first place, its proof depends on a theorem that is true for unitary spaces only, and, in the second place, in a real space the reality of (Ax, x) is automatic, whereas the identity $(Ax, y) = (x, Ay)$ is not necessarily satisfied.

§ 72. Positive transformations

When is a complex number ζ positive (that is, ≥ 0)? Two equally natural necessary and sufficient conditions are that ζ may be written in the form $\zeta = \xi^2$ with some real ξ, or that ζ may be written in the form $\zeta = \bar{\sigma}\sigma$ with some σ (in general complex). Remembering also the fact that (at least for unitary spaces) the Hermitian character of a transformation A can be described in terms of the inner products (Ax, x), we may consider any one of the three conditions below and attempt to use it as the definition of positiveness for transformations:

(1) $\qquad\qquad A = B^2$ for some self-adjoint B,

(2) $\qquad\qquad A = C^*C$ for some C,

(3) $\qquad\qquad A$ is self-adjoint and $(Ax, x) \geqq 0$ for all x.

Before deciding which one of these three conditions to use as definition, we observe that (1) $\Rightarrow$ (2) $\Rightarrow$ (3). Indeed: if $A = B^2$ and $B = B^*$, then $A = BB = B^*B$, and if $A = C^*C$, then $A^* = C^*C = A$ and $(Ax, x) = (C^*Cx, x) = (Cx, Cx) = \| Cx \|^2 \geqq 0$. It is actually true that (3) implies (1), so that the three conditions are equivalent, but we shall not be able to prove this until later. We adopt as our definition the third condition.

DEFINITION. A linear transformation A on an inner product space is *positive*, in symbols $A \geqq 0$, if it is self-adjoint and if $(Ax, x) \geqq 0$ for all x.

More generally, we shall write $A \geqq B$ (or $B \leqq A$) whenever $A - B \geqq 0$. Although, of course, it is quite possible that the difference of two transformations that are not even self-adjoint turns out to be positive, we shall generally write inequalities for self-adjoint transformations only. Observe that for a complex inner product space a part of the definition of positiveness is superfluous; if $(Ax, x) \geqq 0$ for all x, then, in particular, (Ax, x) is real for all x, and, by Theorem 4 of the preceding section, A must be positive.

Positive transformations are usually called *non-negative semidefinite*. If $A \geqq 0$ and $(Ax, x) = 0$ implies that $x = 0$, we shall say that A is *strictly positive*; the usual term is *positive definite*. Since the Schwarz inequality implies that

$$| (Ax, x) | \leqq \| Ax \| \cdot \| x \|,$$

we see that if A is a strictly positive transformation and if $Ax = 0$, then $x = 0$, so that, on a finite-dimensional inner product space, a strictly positive transformation is invertible. We shall see later that the converse is true; if $A \geqq 0$ and A is invertible, then A is strictly positive. It is sometimes convenient to indicate the fact that a transformation A is strictly positive by writing $A > 0$; if $A - B > 0$, we may also write $A > B$ (or $B < A$).

It is possible to give a matricial characterization of positive transformations; we shall postpone this discussion till later. In the meantime we shall have occasion to refer to positive matrices, meaning thereby Hermitian symmetric matrices (α_{ij}) (that is, $\alpha_{ij} = \overline{\alpha_{ji}}$) with the property that for every sequence $(\xi_1, \cdots, \xi_n)$ of n scalars we have $\sum_i \sum_j \alpha_{ij} \xi_i \bar{\xi}_j \geq 0$. (In the real case the bars may be omitted; in the complex case Hermitian symmetry follows from the other condition.) These conditions are clearly equivalent to the condition that (α_{ij}) be the matrix, with respect to some orthonormal coordinate system, of a positive transformation.

The algebraic rules for combining positive transformations are similar to those for self-adjoint transformations as far as sums, scalar multiples, and inverses are concerned; even § 70, Theorem 2, remains valid if we replace "self-adjoint" by "positive" throughout. It is also true that if A and B are positive, then a necessary and sufficient condition that AB (or BA) be positive is that $AB = BA$ (that is, that A and B commute), but we shall have to postpone the proof of this statement for a while.

EXERCISES

1. Under what conditions on a linear transformation A does the function of two variables, whose value at x and y is (Ax, y), satisfy the conditions on an inner product?

2. Which of the following matrices are positive?

(a) $\begin{pmatrix} 1 & 1 & 1 \\ 1 & 1 & 1 \\ 1 & 1 & 1 \end{pmatrix}.$

(b) $\begin{pmatrix} 0 & i \\ -i & 0 \end{pmatrix}.$

(c) $\begin{pmatrix} 0 & 1 \\ -1 & 0 \end{pmatrix}.$

(d) $\begin{pmatrix} 1 & 1 \\ 1 & 0 \end{pmatrix}.$

(e) $\begin{pmatrix} 1 & 1 & 1 \\ 0 & 1 & 1 \\ 0 & 0 & 1 \end{pmatrix}.$

3. For which values of α is the matrix

$$\begin{pmatrix} \alpha & 1 & 1 \\ 1 & 0 & 0 \\ 1 & 0 & 0 \end{pmatrix}$$

positive?

4. (a) If A is self-adjoint, then tr A is real.
(b) If $A \geq 0$, then tr $A \geq 0$.

5. (a) Give an example of a positive matrix some of whose entries are negative.
(b) Give an example of a non-positive matrix all of whose entries are positive.

6. A necessary and sufficient condition that a two-by-two matrix $\begin{pmatrix} \alpha & \beta \\ \gamma & \delta \end{pmatrix}$ (considered as a linear transformation on $\mathbb{C}^2$) be positive is that it be Hermitian sym-

metric (that is, that α and δ be real and $\gamma = \bar{\beta}$) and that $\alpha \geqq 0$, $\delta \geqq 0$, and $\alpha\delta - \beta\gamma \geqq 0$.

7. Associated with each sequence $(x_1, \cdots, x_k)$ of k vectors in an inner product space there is a k-by-k matrix (not a linear transformation) called the *Gramian* of $(x_1, \cdots, x_k)$ and denoted by $G(x_1, \cdots, x_k)$; the element in the i-th row and j-th column of $G(x_1, \cdots, x_k)$ is the inner product (x_i, x_j). Prove that every Gramian is a positive matrix.

8. If x and y are non-zero vectors (in a finite-dimensional inner product space), then a necessary and sufficient condition that there exist a positive transformation A such that $Ax = y$ is that $(x, y) > 0$.

9. (a) If the matrices $A = \begin{pmatrix} 1 & 0 \\ 0 & 0 \end{pmatrix}$ and $B = \begin{pmatrix} 0 & 0 \\ 0 & 1 \end{pmatrix}$ are considered as linear transformations on $\mathbb{C}^2$, and if C is a Hermitian matrix (linear transformation on $\mathbb{C}^2$) such that $A \leqq C$ and $B \leqq C$, then

$$C = \begin{pmatrix} 1 + \epsilon & \theta \\ \theta & 1 + \delta \end{pmatrix},$$

where ϵ and δ are positive real numbers and $|\bar{\theta}|^2 \leqq \min \{\epsilon(1 + \delta), \delta(1 + \epsilon)\}$.

(b) If, moreover, $C \leqq 1$, then $\epsilon = \delta = \theta = 0$. In modern terminology these facts together show that Hermitian matrices with the ordering induced by the notion of positiveness do not form a *lattice*. In the real case, if the matrix $\begin{pmatrix} \alpha & \beta \\ \beta & \gamma \end{pmatrix}$ is interpreted as the point (α, β, γ) in three-dimensional space, the ordering and its non-lattice character take on an amusing geometric aspect.

§ 73. Isometries

We continue with our program of investigating the analogy between numbers and transformations. When does a complex number ζ have absolute value one? Clearly a necessary and sufficient condition is that $\bar{\zeta} = 1/\zeta$; guided by our heuristic principle, we are led to consider linear transformations U for which $U^* = U^{-1}$, or, equivalently, for which $UU^* = U^*U = 1$. (We observe that on a finite-dimensional vector space either of the two conditions $UU^* = 1$ and $U^*U = 1$ implies the other; see § 36, Theorems 1 and 2.) Such transformations are called *orthogonal* or *unitary* according as the underlying inner product space is real or complex. We proceed to derive a couple of useful alternative characterizations of them.

THEOREM. *The following three conditions on a linear transformation U on an inner product space are equivalent to each other.*

(1) $\qquad\qquad\qquad U^*U = 1,$

(2) $\qquad\qquad\qquad (Ux, Uy) = (x, y)$ for all x and y,

(3) $\qquad\qquad\qquad \| Ux \| = \| x \|$ for all x.

PROOF. If (1) holds, then

$$(Ux, Uy) = (U^*Ux, y) = (x, y)$$

for all x and y, and, in particular,

$$\| Ux \|^2 = \| x \|^2$$

for all x; this proves both the implications (1) $\Rightarrow$ (2) and (2) $\Rightarrow$ (3). The proof can be completed by showing that (3) implies (1). If (3) holds, that is, if $(U^*Ux, x) = (x, x)$ for all x, then § 71, Theorem 2 is applicable to the (self-adjoint) transformation $U^*U - 1$; the conclusion is that $U^*U = 1$ (as desired).

Since (3) implies that

(4) $$\| Ux - Uy \| = \| x - y \|$$

for all x and y (the converse implication (4) $\Rightarrow$ (3) is also true and trivial), we see that transformations of the type that the theorem deals with are characterized by the fact that they preserve distances. For this reason we shall call such a transformation an *isometry*. Since, as we have already remarked, an isometry on a finite-dimensional space is necessarily orthogonal or unitary (according as the space is real or complex), use of this terminology will enable us to treat the real and the complex cases simultaneously. We observe that (on a finite-dimensional space) an isometry is always invertible and that U^{-1} ($= U^*$) is an isometry along with U.

In any algebraic system, and in particular in general vector spaces and inner product spaces, it is of interest to consider the automorphisms of the system, that is, to consider those one-to-one mappings of the system onto itself that preserve all the structural relations among its elements. We have already seen that the automorphisms of a general vector space are the invertible linear transformations. In an inner product space we require more of an automorphism, namely, that it also preserve inner products (and consequently lengths and distances). The preceding theorem shows that this requirement is equivalent to the condition that the transformation be an isometry. (We are assuming finite-dimensionality here; on infinite-dimensional spaces the range of an isometry need not be the entire space. This unimportant sacrifice in generality is for the sake of terminological convenience; for infinite-dimensional spaces there is no commonly used word that describes orthogonal and unitary transformations simultaneously.) Thus the two questions "What linear transformations are the analogues of complex numbers of absolute value one?" and "What are the most general automorphisms of a finite-dimensional inner product space?" have the same answer: isometries. In the next section we shall show that isometries also furnish the answer to a third important question.

§ 74. Change of orthonormal basis

We have seen that the theory of the passage from one linear basis of a vector space to another is best studied by means of an associated linear transformation A (§§ 46, 47); the question arises as to what special properties A has when we pass from one *orthonormal* basis of an inner product space to another. The answer is easy.

Theorem 1. *If* $\mathfrak{X} = \{x_1, \cdots, x_n\}$ *is an orthonormal basis of an n-dimensional inner product space* $\mathcal{V}$, *and if* U *is an isometry on* $\mathcal{V}$, *then* $U\mathfrak{X} = \{Ux_1, \cdots, Ux_n\}$ *is also an orthonormal basis of* $\mathcal{V}$. *Conversely, if* U *is a linear transformation and* $\mathfrak{X}$ *is an orthonormal basis with the property that* $U\mathfrak{X}$ *is also an orthonormal basis, then* U *is an isometry.*

proof. Since $(Ux_i, Ux_j) = (x_i, x_j) = \delta_{ij}$, it follows that $U\mathfrak{X}$ is an orthonormal set along with $\mathfrak{X}$; it is complete if $\mathfrak{X}$ is, since $(x, Ux_i) = 0$ for $i = 1, \cdots, n$ implies that $(U^*x, x_i) = 0$ and hence that $U^*x = x = 0$. If, conversely, $U\mathfrak{X}$ is a complete orthonormal set along with $\mathfrak{X}$, then we have $(Ux, Uy) = (x, y)$ whenever x and y are in $\mathfrak{X}$, and it is clear that by linearity we obtain $(Ux, Uy) = (x, y)$ for all x and y.

We observe that the matrix (u_{ij}) of an isometric transformation, with respect to an arbitrary orthonormal basis, satisfies the conditions

$$\sum_k \bar{u}_{ki} u_{kj} = \delta_{ij},$$

and that, conversely, any such matrix, together with an orthonormal basis, defines an isometry. (Proof: $U^*U = 1$. In the real case the bars may be omitted.) For brevity we shall say that a matrix satisfying these conditions is an *isometric matrix*.

An interesting and easy consequence of our considerations concerning isometries is the following corollary of § 56, Theorem 1.

Theorem 2. *If* A *is a linear transformation on a complex n-dimensional inner product space* $\mathcal{V}$, *then there exists an orthonormal basis* $\mathfrak{X}$ *in* $\mathcal{V}$ *such that the matrix* $[A; \mathfrak{X}]$ *is triangular, or equivalently, if* $[A]$ *is a matrix, then there exists an isometric matrix* $[U]$ *such that* $[U]^{-1}[A][U]$ *is triangular.*

proof. In § 56, in the derivation of Theorem 2 from Theorem 1, we constructed a (linear) basis $\mathfrak{X} = \{x_1, \cdots, x_n\}$ with the property that $x_1, \cdots, x_j$ lie in $\mathfrak{M}_j$ and span $\mathfrak{M}_j$ for $j = 1, \cdots, n$, and we showed that with respect to this basis the matrix of A is triangular. If we knew that this basis is also an orthonormal basis, we could apply Theorem 1 of the present section to obtain the desired result. If $\mathfrak{X}$ is not an orthonormal basis, it is easy to make it into one; this is precisely what the Gram-Schmidt orthogonalization process (§ 65) can do. Here we use a special property of the

Gram-Schmidt process, namely, that the j-th element of the orthonormal basis it constructs is a linear combination of $x_1, \cdots, x_j$ and lies therefore in $\mathfrak{M}_j$.

1. If $(Ax)(t) = x(-t)$ on $\mathcal{O}$ (with the inner product given by $(x, y) = \int_0^1 x(t)\overline{y(t)}\,dt$) is the linear transformation A isometric? Is it self-adjoint?

2. For which values of α are the following matrices isometric?

(a) $\begin{pmatrix} \alpha & 0 \\ 1 & 1 \end{pmatrix}.$ (b) $\begin{pmatrix} \alpha & \frac{1}{2} \\ -\frac{1}{2} & \alpha \end{pmatrix}.$

3. Find a 3-by-3 isometric matrix whose first row is a multiple of $(1, 1, 1)$.

4. If a linear transformation has any two of the properties of being self-adjoint, isometric, or involutory, then it has the third. (Recall that an involution is a linear transformation A such that $A^2 = 1$.)

5. If an isometric matrix is triangular, then it is diagonal.

6. If $(x_1, \cdots, x_k)$ and $(y_1, \cdots, y_k)$ are two sequences of vectors in the same inner product space, then a necessary and sufficient condition that there exist an isometry U such that $Ux_i = y_i, i = 1, \cdots, k$, is that $(x_1, \cdots, x_k)$ and $(y_1, \cdots, y_k)$ have the same Gramian.

7. The mapping $\xi \to \dfrac{\xi + 1}{\xi - 1}$ maps the imaginary axis in the complex plane once around the unit circle, missing the point 1; the inverse mapping (from the circle minus a point to the imaginary axis) is given by the same formula. The transformation analogues of these geometric facts are as follows.
   (a) If A is skew, then $A - 1$ is invertible.
   (b) If $U = (A + 1)(A - 1)^{-1}$, then U is isometric. (Hint: $\| (A + 1)y \|^2 = \| (A - 1)y \|^2$ for every y.)
   (c) $U - 1$ is invertible.
   (d) If U is isometric and $U - 1$ is invertible, and if $A = (U + 1)(U - 1)^{-1}$, then A is skew.
   Each of A and U is known as the *Cayley transform* of the other.

8. Suppose that U is a transformation (not assumed to be linear) that maps an inner product space $\mathcal{V}$ onto itself (that is, if x is in $\mathcal{V}$, then Ux is in $\mathcal{V}$, and if y is in $\mathcal{V}$, then $y = Ux$ for some x in $\mathcal{V}$), in such a way that $(Ux, Uy) = (x, y)$ for all x and y.
   (a) Prove that U is one-to-one and that if the inverse transformation is denoted by U^{-1}, then $(U^{-1}x, U^{-1}y) = (x, y)$ and $(Ux, y) = (x, U^{-1}y)$ for all x and y.
   (b) Prove that U is linear. (Hint: $(x, U^{-1}y)$ depends linearly on x.)

9. A *conjugation* is a transformation J (not assumed to be linear) that maps a unitary space onto itself and is such that $J^2 = 1$ and $(Jx, Jy) = (y, x)$ for all x and y.
   (a) Give an example of a conjugation.
   (b) Prove that $(Jx, y) = (Jy, x)$.
   (c) Prove that $J(x + y) = Jx + Jy$.
   (d) Prove that $J(\alpha x) = \bar{\alpha} \cdot Jx$.

10. A linear transformation A is said to be *real* with respect to a conjugation J if $AJ = JA$.

(a) Give an example of a Hermitian transformation that is not real, and give an example of a real transformation that is not Hermitian.

(b) If A is real, then the spectrum of A is symmetric about the real axis.

(c) If A is real, then so is A^*.

11. § 74, Theorem 2 shows that the triangular form can be achieved by an orthonormal basis; is the same thing true for the Jordan form?

12. If $\operatorname{tr} A = 0$, then there exists an isometric matrix U such that all the diagonal entries of $[U]^{-1}[A][U]$ are zero. (Hint: see § 56, Ex. 6.)

§ 75. Perpendicular projections

We are now in a position to fulfill our earlier promise to investigate the projections associated with the particular direct sum decompositions $\mathcal{V} = \mathfrak{M} \oplus \mathfrak{M}^\perp$. We shall call such a projection a *perpendicular projection*. Since $\mathfrak{M}^\perp$ is uniquely determined by the subspace $\mathfrak{M}$, we need not specify both the direct summands associated with a projection if we already know that it is perpendicular. We shall call the (perpendicular) projection E on $\mathfrak{M}$ along $\mathfrak{M}^\perp$ simply the projection on $\mathfrak{M}$ and we shall write $E = P_{\mathfrak{M}}$.

THEOREM 1. *A linear transformation E is a perpendicular projection if and only if $E = E^2 = E^*$. Perpendicular projections are positive linear transformations and have the property that $\| Ex \| \leq \| x \|$ for all x.*

PROOF. If E is a perpendicular projection, then § 45, Theorem 1 and the theorem of § 20 show (after, of course, the usual replacements, such as $\mathfrak{M}^\perp$ for $\mathfrak{M}^0$ and A^* for A') that $E = E^*$. Conversely if $E = E^2 = E^*$, then the idempotence of E assures us that E is the projection on $\mathfrak{R}$ along $\mathfrak{N}$, where, of course, $\mathfrak{R} = \mathfrak{R}(E)$ and $\mathfrak{N} = \mathfrak{N}(E)$ are the range and the null-space of E, respectively. Hence we need only show that $\mathfrak{R}$ and $\mathfrak{N}$ are orthogonal. For this purpose let x be any element of $\mathfrak{R}$ and y any element of $\mathfrak{N}$; the desired result follows from the relation

$$(x, y) = (Ex, y) = (x, E^*y) = (x, Ey) = 0.$$

The positive character of an E satisfying $E = E^2 = E^*$ follows from

$$(Ex, x) = (E^2x, x) = (Ex, E^*x) = (Ex, Ex) = \| Ex \|^2 \geq 0.$$

Applying this result to the perpendicular projection $1 - E$, we see that

$$\| x \|^2 - \| Ex \|^2 = (x, x) - (Ex, x) = ([1 - E]x, x) \geq 0;$$

this concludes the proof of the theorem.

For some of the generalizations of our theory it is useful to know that idempotence together with the last property mentioned in Theorem 1 is also characteristic of perpendicular projections.

Theorem 2. *If a linear transformation E is such that $E = E^2$ and $\| Ex \| \leq \| x \|$ for all x, then $E = E^*$.*

proof. We are to show that the range $\mathfrak{R}$ and the null-space $\mathfrak{N}$ of E are orthogonal. If x is in $\mathfrak{N}^{\perp}$, then $y = Ex - x$ is in $\mathfrak{N}$, since $Ey = E^2x - Ex = Ex - Ex = 0$. Hence $Ex = x + y$ with $(x, y) = 0$, so that

$$\| x \|^2 \geq \| Ex \|^2 = \| x \|^2 + \| y \|^2 \geq \| x \|^2,$$

and therefore $y = 0$. Consequently $Ex = x$, so that x is in $\mathfrak{R}$; this proves that $\mathfrak{N}^{\perp} \subset \mathfrak{R}$. Conversely, if z is in $\mathfrak{R}$, so that $Ez = z$, we write $z = x + y$ with x in $\mathfrak{N}^{\perp}$ and y in $\mathfrak{N}$. Then $z = Ez = Ex + Ey = Ex = x$. (The reason for the last equality is that x is in $\mathfrak{N}^{\perp}$ and therefore in $\mathfrak{R}$.) Hence z is in $\mathfrak{N}^{\perp}$, so that $\mathfrak{R} \subset \mathfrak{N}^{\perp}$, and therefore $\mathfrak{R} = \mathfrak{N}^{\perp}$.

We shall need also the fact that the theorem of § 42 remains true if the word "projection" is qualified throughout by "perpendicular." This is an immediate consequence of the preceding characterization of perpendicular projections and of the fact that sums and differences of self-adjoint transformations are self-adjoint, whereas the product of two self-adjoint transformations is self-adjoint if and only if they commute. By our present geometric methods it is also quite easy to generalize the part of the theorem dealing with sums from two summands to any finite number. The generalization is most conveniently stated in terms of the concept of orthogonality for projections; we shall say that two (perpendicular) projections E and F are *orthogonal* if $EF = 0$. (Consideration of adjoints shows that this is equivalent to $FE = 0$.) The following theorem shows that the geometric language is justified.

Theorem 3. *Two perpendicular projections $E = P_{\mathfrak{M}}$ and $F = P_{\mathfrak{N}}$ are orthogonal if and only if the subspaces $\mathfrak{M}$ and $\mathfrak{N}$ (that is, the ranges of E and F) are orthogonal.*

proof. If $EF = 0$, and if x and y are in the ranges of E and F respectively, then

$$(x, y) = (Ex, Fy) = (x, E^*Fy) = (x, EFy) = 0.$$

If, conversely, $\mathfrak{M}$ and $\mathfrak{N}$ are orthogonal (so that $\mathfrak{N} \subset \mathfrak{M}^{\perp}$), then the fact that $Ex = 0$ for x in $\mathfrak{M}^{\perp}$ implies that $EFx = 0$ for all x (since Fx is in $\mathfrak{N}$ and consequently in $\mathfrak{M}^{\perp}$).

§ 76. Combinations of perpendicular projections

The sum theorem for perpendicular projections is now easy.

THEOREM 1. *If $E_1, \cdots, E_n$ are (perpendicular) projections, then a necessary and sufficient condition that $E = E_1 + \cdots + E_n$ be a (perpendicular) projection is that $E_i E_j = 0$ whenever $i \neq j$ (that is, that the E_i be pairwise orthogonal).*

PROOF. The proof of the sufficiency of the condition is trivial; we prove explicitly its necessity only, so that we now assume that E is a perpendicular projection. If x belongs to the range of some E_i, then

$$\| x \|^2 \geq \| Ex \|^2 = (Ex, x) = (\textstyle\sum_j E_j x, x)$$
$$= \textstyle\sum_j (E_j x, x) = \textstyle\sum_j \| E_j x \|^2 \geq \| E_i x \|^2 = \| x \|^2,$$

so that we must have equality all along. Since, in particular, we must have

$$\textstyle\sum_j \| E_j x \|^2 = \| E_i x \|^2,$$

it follows that $E_j x = 0$ whenever $j \neq i$. In other words, every x in the range of E_i is in the null-space (and, consequently, is orthogonal to the range) of every E_j with $j \neq i$; using § 75, Theorem 3, we draw the desired conclusion.

We end our discussion of projections with a brief study of order relations. It is tempting to write $E \leq F$, for two perpendicular projections $E = P_{\mathfrak{M}}$ and $F = P_{\mathfrak{N}}$, whenever $\mathfrak{M} \subset \mathfrak{N}$. Earlier, however, we interpreted the sign $\leq$, when used in an expression involving linear transformations E and F (as in $E \leq F$), to mean that $F - E$ is a positive transformation. There are also other possible reasons for considering E to be smaller than F; we might have $\| Ex \| \leq \| Fx \|$ for all x, or $FE = EF = E$ (see § 42, (ii)). The situation is straightened out by the following theorem, which plays here a role similar to that of § 75, Theorem 3, that is, it establishes the coincidence of several seemingly different concepts concerning projections, some of which are defined algebraically while others refer to the underlying geometrical objects.

THEOREM 2. *For perpendicular projections $E = P_{\mathfrak{M}}$ and $F = P_{\mathfrak{N}}$ the following conditions are mutually equivalent.*

(i) $\qquad\qquad\qquad\qquad E \leq F.$

(ii) $\qquad\qquad\qquad \| Ex \| \leq \| Fx \|$ for all x.

(iii) $\qquad\qquad\qquad\qquad \mathfrak{M} \subset \mathfrak{N}.$

(iva) $\qquad\qquad\qquad\qquad FE = E,$

(ivb) $\qquad\qquad\qquad\qquad EF = E.$

PROOF. We shall prove the implication relations (i) $\Rightarrow$ (ii) $\Rightarrow$ (iii) $\Rightarrow$ (iva) $\Rightarrow$ (ivb) $\Rightarrow$ (i).

(i) $\Rightarrow$ (ii). If $E \leq F$, then, for all x,

$$0 \leq ([F - E]x, x) = (Fx, x) - (Ex, x) = \| Fx \|^2 - \| Ex \|^2$$

(since E and F are perpendicular projections).

(ii) $\Rightarrow$ (iii). We assume that $\| Ex \| \leq \| Fx \|$ for all x. Let us now take any x in $\mathfrak{M}$; then we have

$$\| x \| \geq \| Fx \| \geq \| Ex \| = \| x \|,$$

so that $\| Fx \| = \| x \|$, or $(x, x) - (Fx, x) = 0$, whence

$$([1 - F]x, x) = \| (1 - F)x \|^2 = 0,$$

and consequently $x = Fx$. In other words, x in $\mathfrak{M}$ implies that x is in $\mathfrak{N}$, as was to be proved.

(iii) $\Rightarrow$ (iva). If $\mathfrak{M} \subset \mathfrak{N}$, then Ex is in $\mathfrak{N}$ for all x, so that, $FEx = Ex$ for all x, as was to be proved.

That (iva) implies (ivb), and is in fact equivalent to it, follows by taking adjoints.

(iv) $\Rightarrow$ (i). If $EF = FE = E$, then, for all x,

$$(Fx, x) - (Ex, x) = (Fx, x) - (FEx, x) = (F[1 - E]x, x).$$

Since E and F are commutative projections, so also are $(1 - E)$ and F, and consequently $G = F(1 - E)$ is a projection. Hence

$$(Fx, x) - (Ex, x) = (Gx, x) = \| Gx \|^2 \geq 0.$$

This completes the proof of Theorem 2.

In terms of the concepts introduced by now, it is possible to give a quite intuitive sounding formulation of the theorem of § 42 (in so far as it applies to perpendicular projections), as follows. For two perpendicular projections E and F, their sum, product, or difference is also a perpendicular projection if and only if F is respectively orthogonal to, commutative with, or greater than E.

EXERCISES

1. (a) Give an example of a projection that is not a perpendicular projection.

(b) Give an example of two projections E and F (they cannot both be perpendicular) such that $EF = 0$ and $FE \neq 0$.

2. Find the (perpendicular) projection of $(1, 1, 1)$ on the (one-dimensional) subspace of $\mathbb{C}^3$ spanned by $(1, -1, 1)$. (In other words: find the image of the given vector under the projection onto the given subspace.)

3. Find the matrices of all perpendicular projections on $\mathbb{C}^2$.

4. If $U = 2E - 1$, then a necessary and sufficient condition that U be an involutory isometry is that E be a perpendicular projection.

5. A linear transformation U is called a *partial isometry* if there exists a subspace $\mathfrak{M}$ such that $\| Ux \| = \| x \|$ whenever x is in $\mathfrak{M}$ and $Ux = 0$ whenever x is in $\mathfrak{M}^{\perp}$.

(a) The adjoint of a partial isometry is a partial isometry.

(b) If U is a partial isometry and if $\mathfrak{M}$ is a subspace such that $\| Ux \| = \| x \|$ or 0 according as x is in $\mathfrak{M}$ or in $\mathfrak{M}^{\perp}$, then U^*U is the perpendicular projection on $\mathfrak{M}$.

(c) Each of the following four conditions is necessary and sufficient that a linear transformation U be a partial isometry. (i) $UU^*U = U$, (ii) U^*U is a projection, (iii) $U^*UU^* = U^*$, (iv) UU^* is a projection.

(d) If λ is a proper value of a partial isometry, then $|\lambda| \leqq 1$.

(e) Give an example of a partial isometry that has $\frac{1}{2}$ as a proper value.

6. Suppose that A is a linear transformation on, and $\mathfrak{M}$ is a subspace of, a finite-dimensional vector space $\mathcal{V}$. Prove that if dim $\mathfrak{M} \leqq$ dim $\mathfrak{M}^{\perp}$, then there exist linear transformations B and C on $\mathcal{V}$ such that $Ax = (BC - CB)x$ for all x in $\mathfrak{M}$. (Hint: let B be a partial isometry such that $\| Bx \| = \| x \|$ or 0 according as x is in $\mathfrak{M}$ or in $\mathfrak{M}^{\perp}$ and such that $\mathcal{R}(B) \subset \mathfrak{M}^{\perp}$.)

§ 77. Complexification

In the past few sections we have been treating real and complex vector spaces simultaneously. Sometimes this is not possible; the complex number system is richer than the real. There are theorems that are true for both real and complex spaces, but for which the proof is much easier in the complex case, and there are theorems that are true for complex spaces but not for real ones. (An example of the latter kind is the assertion that if the space is finite-dimensional, then every linear transformation has a proper value.) For these reasons, it is frequently handy to be able to "complexify" a real vector space, that is, to associate with it a complex vector space with essentially the same properties. The purpose of this section is to describe such a process of complexification.

Suppose that $\mathcal{V}$ is a real vector space, and let $\mathcal{V}^+$ be the set of all ordered pairs $\langle x, y \rangle$ with both x and y in $\mathcal{V}$. Define the sum of two elements of $\mathcal{V}^+$ by

$$\langle x_1, y_1 \rangle + \langle x_2, y_2 \rangle = \langle x_1 + x_2, y_1 + y_2 \rangle,$$

and define the product of an element of $\mathcal{V}^+$ by a complex number $\alpha + i\beta$ (α and β real, $i = \sqrt{-1}$) by

$$(\alpha + i\beta)\langle x, y \rangle = \langle \alpha x - \beta y, \beta x + \alpha y \rangle.$$

(To remember these formulas, pretend that $\langle x, y \rangle$ means $x + iy$.) A straightforward and only slightly laborious computation shows that the set $\mathcal{U}^+$ becomes a complex vector space with respect to these definitions of the linear operations.

The set of those elements $\langle x, y \rangle$ of $\mathcal{U}^+$ for which $y = 0$ is in a natural one-to-one correspondence with the space $\mathcal{U}$. Being a complex vector space, the space $\mathcal{U}^+$ may also be regarded as a real vector space; if we identify each element x of $\mathcal{U}$ with its replica $\langle x, 0 \rangle$ in $\mathcal{U}^+$ (it is exceedingly convenient to do this), we may say that $\mathcal{U}^+$ (as a real vector space) includes $\mathcal{U}$. Since $\langle 0, y \rangle = i\langle y, 0 \rangle$, so that $\langle x, y \rangle = \langle x, 0 \rangle + i\langle y, 0 \rangle$, our identification convention enables us to say that every vector in $\mathcal{U}^+$ has the form $x + iy$, with x and y in $\mathcal{U}$. Since $\mathcal{U}$ and $i\mathcal{U}$ (where $i\mathcal{U}$ denotes the set of all elements $\langle x, y \rangle$ in $\mathcal{U}^+$ with $x = 0$) are subsets of $\mathcal{U}^+$ with only 0 (that is, $\langle 0, 0 \rangle$) in common, it follows that the representation of a vector of $\mathcal{U}^+$ in the form $x + iy$ (with x and y in $\mathcal{U}$) is unique. We have thus constructed a complex vector space $\mathcal{U}^+$ with the property that $\mathcal{U}^+$ considered as a real space includes $\mathcal{U}$ as a subspace, and such that $\mathcal{U}^+$ is the direct sum of $\mathcal{U}$ and $i\mathcal{U}$. (Here $i\mathcal{U}$ denotes the set of all those elements of $\mathcal{U}^+$ that have the form iy for some y in $\mathcal{U}$.) We shall call $\mathcal{U}^+$ the *complexification* of $\mathcal{U}$.

If $\{x_1, \cdots, x_n\}$ is a linearly independent set in $\mathcal{U}$ (real coefficients), then it is also a linearly independent set in $\mathcal{U}^+$ (complex coefficients). Indeed, if $\alpha_1, \cdots, \alpha_n, \beta_1, \cdots, \beta_n$ are real numbers such that $\sum_j (\alpha_j + i\beta_j)x_j = 0$, then $(\sum_j \alpha_j x_j) + i(\sum_j \beta_j x_j) = 0$, and consequently, by the uniqueness of the representation of vectors in $\mathcal{U}^+$ by means of vectors in $\mathcal{U}$, it follows that $\sum_j \alpha_j x_j = \sum_j \beta_j x_j = 0$; the desired result is now implied by the assumed (real) linear independence of $\{x_1, \cdots, x_n\}$ in $\mathcal{U}$. If, moreover, $\{x_1, \cdots, x_n\}$ is a basis in $\mathcal{U}$ (real coefficients), then it is also a basis in $\mathcal{U}^+$ (complex coefficients). Indeed, if x and y are in $\mathcal{U}$, then there exist real numbers $\alpha_1, \cdots, \alpha_n, \beta_1, \cdots, \beta_n$ such that $x = \sum_j \alpha_j x_j$ and $y = \sum_j \beta_j x_j$; it follows that $x + iy = \sum_j (\alpha_j + i\beta_j)x_j$, and hence that $\{x_1, \cdots, x_n\}$ spans $\mathcal{U}^+$. These results imply that the complex vector space $\mathcal{U}^+$ has the same dimension as the real vector space $\mathcal{U}$.

There is a natural way to extend every linear transformation A on $\mathcal{U}$ to a linear transformation A^+ on $\mathcal{U}^+$; we write

$$A^+(x + iy) = Ax + iAy$$

whenever x and y are in $\mathcal{U}$. (The verification that A^+ is indeed a linear transformation on $\mathcal{U}^+$ is routine.) A similar extension works for linear and even multilinear functionals. If, for instance, w is a (real) bilinear functional on $\mathcal{U}$, its extension to $\mathcal{U}^+$ is the (complex) bilinear functional defined by

$$w^+(x_1 + iy_1, x_2 + iy_2)$$

$$= w(x_1, x_2) - w(y_1, y_2) + i(w(x_1, y_2) + w(y_1, x_2)).$$

If, on the other hand, w is alternating, then the same is true of w^+. Indeed, the real and imaginary parts of $w^+(x + iy, x + iy)$ are $w(x, x) - w(y, y)$ and $w(x, y) + w(y, x)$ respectively; if w is alternating, then w is skew symmetric (§ 30, Theorem 1), and therefore w^+ is alternating. The same proof establishes the corresponding result for k-linear functionals also, for all values of k. From this and from the definition of determinants it follows that $\det A = \det A^+$ for every linear transformation A on $\mathcal{V}$.

The method of extending bilinear functionals works for conjugate bilinear functionals also. If, that is, $\mathcal{V}$ is a (real) inner product space, then there is a natural way of introducing a (complex) inner product into $\mathcal{V}^+$; we write, by definition,

$$(x_1 + iy_1, x_2 + iy_2) = (x_1, x_2) + (y_1, y_2) - i((x_1, y_2) - (y_1, x_2)).$$

Observe that if x and y are orthogonal vectors in $\mathcal{V}$, then

$$\| x + iy \|^2 = \| x \|^2 + \| y \|^2.$$

The correspondence from A to A^+ preserves all algebraic properties of transformations. Thus if $B = \alpha A$ (with α real), then $B^+ = \alpha A^+$; if $C = A + B$, then $C^+ = A^+ + B^+$; and if $C = AB$, then $C^+ = A^+B^+$. If, moreover, $\mathcal{V}$ is an inner product space, and if $B = A^*$, then $B^+ = (A^+)^*$. (Proof: evaluate $(A^+(x_1 + iy_1), (x_2 + iy_2))$ and $(x_1 + iy_1, B^+(x_2 + iy_2))$.)

If A is a linear transformation on $\mathcal{V}$ and if A^+ has a proper vector $x + iy$, with proper value $\alpha + i\beta$ (where x and y are in $\mathcal{V}$ and α and β are real), so that

$$Ax = \alpha x - \beta y,$$

$$Ay = \beta x + \alpha y,$$

then the subspace of $\mathcal{V}$ spanned by x and y is invariant under A. (Since every linear transformation on a complex vector space has a proper vector, we conclude that every linear transformation on a real vector space leaves invariant a subspace of dimension equal to 1 or 2.) If, in particular, A^+ happens to have a real proper value (that is, if $\beta = 0$), then A has the same proper value (since $Ax = \alpha x$, $Ay = \alpha y$, and not both x and y can vanish).

We have already seen that every (real) basis in $\mathcal{V}$ is at the same time a (complex) basis in $\mathcal{V}^+$. It follows that the matrix of a linear transformation A on $\mathcal{V}$, with respect to some basis $\mathfrak{X}$ in $\mathcal{V}$, is the same as the matrix of A^+ on $\mathcal{V}^+$, with respect to the basis $\mathfrak{X}$ in $\mathcal{V}^+$. This comment is at the root of the whole theory of complexification; the naive point of view on the matter is that real matrices constitute a special case of complex matrices.

1. What happens if the process of complexification described in § 77 is applied to a vector space that is already complex?

2. Prove that there exists a unique isomorphism between the complexification described in § 77 and the one described in § 25, Ex. 5 with the property that each "real" vector (that is, each vector in the originally given real vector space) corresponds to itself.

3. (a) What is the complexification of $\mathcal{R}^1$?
(b) If $\mathcal{V}$ is an n-dimensional real vector space, what is the dimension of its complexification $\mathcal{V}^+$, regarded as a real vector space?

4. Suppose that $\mathcal{V}^+$ is the complex inner product space obtained by complexifying a real inner product space $\mathcal{V}$.
(a) Prove that if $\mathcal{V}^+$ is regarded as a real vector space and if $A(x + iy) = x - iy$ whenever x and y are in $\mathcal{V}$, then A is a linear transformation on $\mathcal{V}^+$.
(b) Is A self-adjoint? Isometric? Idempotent? Involutory?
(c) What if $\mathcal{V}^+$ is regarded as a complex space?

5. Discuss the relation between duality and complexification, and, in particular, the relation between the adjoint of a linear transformation on a real vector space and the adjoint of its complexification.

6. If A is a linear transformation on a real vector space $\mathcal{V}$ and if a subspace $\mathfrak{M}$ of the complexification $\mathcal{V}^+$ is invariant under A^+, then $\mathfrak{M}^\perp \cap \mathcal{V}$ is invariant under A.

§ 78. Characterization of spectra

The following results support the analogy between numbers and transformations more than anything so far; they assert that the properties that caused us to define the special classes of transformations we have been considering are reflected by their spectra.

THEOREM 1. *If A is a self-adjoint transformation on an inner product space, then every proper value of A is real; if A is positive, or strictly positive, then every proper value of A is positive, or strictly positive, respectively.*

PROOF. We may ignore the fact that the first assertion is trivial in the real case; the same proof serves to establish both assertions in both the real and the complex case. Indeed, if $Ax = \lambda x$, with $x \neq 0$, then,

$$\frac{(Ax, x)}{\| x \|^2} = \frac{\lambda(x, x)}{\| x \|^2} = \lambda;$$

it follows that if (Ax, x) is real (see § 71, Theorem 4), then so is λ, and if (Ax, x) is positive (or strictly positive) then so is λ.

THEOREM 2. *Every root of the characteristic equation of a self-adjoint transformation on a finite-dimensional inner product space is real.*

PROOF. In the complex case roots of the characteristic equation are the same thing as proper values, and the result follows from Theorem 1. If A is a symmetric transformation on a Euclidean space, then its complexification A^+ is Hermitian, and the result follows from the fact that A and A^+ have the same characteristic equation.

We observe that it is an immediate consequence of Theorem 2 that a self-adjoint transformation on a finite-dimensional inner product space always has a proper value.

THEOREM 3. *Every proper value of an isometry has absolute value one.*

PROOF. If U is an isometry, and if $Ux = \lambda x$, with $x \neq 0$, then $\| x \| = \| Ux \| = |\lambda| \cdot \| x \|$.

THEOREM 4. *If A is either self-adjoint or isometric, then proper vectors of A belonging to distinct proper values are orthogonal.*

PROOF. Suppose $Ax_1 = \lambda_1 x_1$, $Ax_2 = \lambda_2 x_2$, $\lambda_1 \neq \lambda_2$. If A is self-adjoint, then

$$(1) \qquad \lambda_1(x_1, x_2) = (Ax_1, x_2) = (x_1, Ax_2) = \lambda_2(x_1, x_2).$$

(The middle step makes use of the self-adjoint character of A, and the last step of the reality of λ_2.) In case A is an isometry, (1) is replaced by

$$(2) \qquad (x_1, x_2) = (Ax_1, Ax_2) = (\lambda_1/\lambda_2)(x_1, x_2);$$

recall that $\bar{\lambda}_2 = 1/\lambda_2$. In either case $(x_1, x_2) \neq 0$ would imply that $\lambda_1 = \lambda_2$, so that we must have $(x_1, x_2) = 0$.

THEOREM 5. *If a subspace $\mathfrak{M}$ is invariant under an isometry U on a finite-dimensional inner product space, then so is $\mathfrak{M}^{\perp}$.*

PROOF. Considered on the finite-dimensional subspace $\mathfrak{M}$, the transformation U is still an isometry, and, consequently, it is invertible. It follows that every x in $\mathfrak{M}$ may be written in the form $x = Uy$ with y in $\mathfrak{M}$; in other words, if x is in $\mathfrak{M}$ and if $y = U^{-1}x$, then y is in $\mathfrak{M}$. Hence $\mathfrak{M}$ is invariant under $U^{-1} = U^*$. It follows from § 45, Theorem 2, that $\mathfrak{M}^{\perp}$ is invariant under $(U^*)^* = U$.

We observe that the same result for self-adjoint transformations (even in not necessarily finite-dimensional spaces) is trivial, since if $\mathfrak{M}$ is invariant under A, then $\mathfrak{M}^{\perp}$ is invariant under $A^* = A$.

THEOREM 6. *If A is a self-adjoint transformation on a finite-dimensional inner product space, then the algebraic multiplicity of each proper value*

λ_0 of A is equal to its geometric multiplicity, that is, to the dimension of the subspace $\mathfrak{M}$ of all solutions of $Ax = \lambda_0 x$.

PROOF. It is clear that $\mathfrak{M}$ is invariant under A, and therefore so is $\mathfrak{M}^{\perp}$; let us denote by B and C the linear transformation A considered only on $\mathfrak{M}$ and $\mathfrak{M}^{\perp}$ respectively. We have

$$\det (A - \lambda) = \det (B - \lambda) \cdot \det (C - \lambda)$$

for all λ. Since B is a self-adjoint transformation on a finite-dimensional space, with only one proper value, namely, λ_0, it follows that λ_0 must occur as a proper value of B with algebraic multiplicity equal to the dimension of $\mathfrak{M}$. If that dimension is m, then $\det (B - \lambda) = (\lambda_0 - \lambda)^m$. Since, on the other hand, λ_0 is not a proper value of C at all, and since, consequently, $\det (C - \lambda_0) \neq 0$, we see that $\det (A - \lambda)$ contains $(\lambda_0 - \lambda)$ as a factor exactly m times, as was to be proved.

What made this proof work was the invariance of $\mathfrak{M}^{\perp}$ and the fact that every root of the characteristic equation of A is a proper value of A. The latter assertion is true for every linear transformation on a unitary space; the following result is a consequence of these observations and of Theorem 5.

THEOREM 7. *If U is a unitary transformation on a finite-dimensional unitary space, then the algebraic multiplicity of each proper value of U is equal to its geometric multiplicity.*

EXERCISES

1. Give an example of a linear transformation with two non-orthogonal proper vectors belonging to distinct proper values.

2. Give an example of a non-positive linear transformation (on a finite-dimensional unitary space) all of whose proper values are positive.

3. (a) If A is self-adjoint, then $\det A$ is real.
(b) If A is unitary, then $|\det A| = 1$.
(c) What can be said about the determinant of a partial isometry?

§ 79. Spectral theorem

We are now ready to prove the main theorem of this book, the theorem of which many of the other results of this chapter are immediate corollaries. To some extent what we have been doing up to now was a matter of sport (useful, however, for generalizations); we wanted to show how much can conveniently be done with spectral theory before proving the spectral theorem. In the complex case, incidentally, the spectral theorem can be made to follow from the triangularization process we have already described; because of the importance of the theorem we prefer to give below its (quite easy) direct proof. The reader may find it profitable to adapt the method of proof (not the result) of § 56, Theorem 2, to prove as much as he can of the spectral theorem and its consequences.

Theorem 1. *To every self-adjoint linear transformation A on a finite-dimensional inner product space there correspond real numbers $\alpha_1, \cdots, \alpha_r$ and perpendicular projections $E_1, \cdots, E_r$ (where r is a strictly positive integer, not greater than the dimension of the space) so that*

(1) *the α_j are pairwise distinct,*
(2) *the E_j are pairwise orthogonal and different from 0,*
(3) $\sum_j E_j = 1$,
(4) $\sum_j \alpha_j E_j = A$.

proof. Let $\alpha_1, \cdots, \alpha_r$ be the distinct proper values of A, and let E_j be the perpendicular projection on the subspace consisting of all solutions of $Ax = \alpha_j x$ $(j = 1, \cdots, r)$. Condition (1) is then satisfied by definition; the fact that the α's are real follows from § 78, Theorem 1. Condition (2) follows from § 78, Theorem 4. From the orthogonality of the E_j we infer that if $E = \sum_j E_j$, then E is a perpendicular projection. The dimension of the range of E is the sum of the dimensions of the ranges of the E_j, and consequently, by § 78, Theorem 6, the dimension of the range of E is equal to the dimension of the entire space; this implies (3). (Alternatively, if $E \neq 1$, then A considered on the range of $1 - E$ would be a self-adjoint transformation with no proper values.) To prove (4), take any vector x and write $x_j = E_j x$; it follows that $Ax_j = \alpha_j x_j$ and hence that

$$Ax = A\left(\sum_j E_j x\right) = \sum_j Ax_j = \sum_j \alpha_j x_j = \sum_j \alpha_j E_j x.$$

This completes the proof of the spectral theorem.

The representation $A = \sum_j \alpha_j E_j$ (where the α's and the E's satisfy the conditions (1)–(3) of Theorem 1) is called a *spectral form* of A; the main effect of the following result is to prove the uniqueness of the spectral form.

Theorem 2. *If $\sum_{j=1}^r \alpha_j E_j$ is the spectral form of a self-adjoint transformation A on a finite-dimensional inner product space, then the α's are all the distinct proper values of A. If, moreover, $1 \leq k \leq r$, then there exist polynomials p_k, with real coefficients, such that $p_k(\alpha_j) = 0$ whenever $j \neq k$ and such that $p_k(\alpha_k) = 1$; for every such polynomial $p_k(A) = E_k$.*

proof. Since $E_j \neq 0$, there exists a vector x in the range of E_j. Since $E_j x = x$ and $E_i x = 0$ whenever $i \neq j$, it follows that

$$Ax = \sum_i \alpha_i E_i x = \alpha_j E_j x = \alpha_j x,$$

so that each α_j is a proper value of A. If, conversely, λ is any proper value of A, say $Ax = \lambda x$ with $x \neq 0$, then we write $x_j = E_j x$ and we see that

$$Ax = \lambda x = \lambda \sum_j x_j$$

and

$$Ax = A \sum_j x_j = \sum_j \alpha_j x_j,$$

so that $\sum_j (\lambda - \alpha_j)x_j = 0$. Since the x_j are pairwise orthogonal, those among them that are not zero form a linearly independent set. It follows that, for each j, either $x_j = 0$ or else $\lambda = \alpha_j$. Since $x \neq 0$, we must have $x_j \neq 0$ for some j, and consequently λ is indeed equal to one of the α's.

Since $E_i E_j = 0$ if $i \neq j$, and $E_j^2 = E_j$, it follows that

$$A^2 = (\sum_i \alpha_i E_i)(\sum_j \alpha_j E_j) = \sum_i \sum_j \alpha_i \alpha_j E_i E_j$$
$$= \sum_j \alpha_j^2 E_j.$$

Similarly

$$A^n = \sum_j \alpha_j^n E_j$$

for every positive integer n (in case $n = 0$, use (3)), and hence

$$p(A) = \sum_j p(\alpha_j)E_j$$

for every polynomial p. To conclude the proof of the theorem, all we need to do is to exhibit a (real) polynomial p_k such that $p_k(\alpha_j) = 0$ whenever $j \neq k$ and such that $p_k(\alpha_k) = 1$. If we write

$$p_k(t) = \prod_{j \neq k} \frac{t - \alpha_j}{\alpha_k - \alpha_j},$$

then p_k is a polynomial with all the required properties.

THEOREM 3. *If $\sum_{j=1}^{r} \alpha_j E_j$ is the spectral form of a self-adjoint transformation A on a finite-dimensional inner product space, then a necessary and sufficient condition that a linear transformation B commute with A is that it commute with each E_j.*

PROOF. The sufficiency of the condition is trivial; if $A = \sum_j \alpha_j E_j$ and $E_j B = B E_j$ for all j, then $AB = BA$. Necessity follows from Theorem 2; if B commutes with A, then B commutes with every polynomial in A, and therefore B commutes with each E_j.

Before exploiting the spectral theorem any further, we remark on its matricial interpretation. If we choose an orthonormal basis in the range of each E_j, then the totality of the vectors in these little bases is a basis for the whole space; expressed in this basis the matrix of A will be diagonal. The fact that by a suitable choice of an orthonormal basis the matrix of a self-adjoint transformation can be made diagonal, or, equivalently, that any self-adjoint matrix can be isometrically transformed (that is, replaced by $[U]^{-1}[A][U]$, where U is an isometry) into a diagonal matrix, already follows (in the complex case) from the theory of the triangular form. We gave the algebraic version for two reasons. First, it is this version that generalizes easily to the infinite-dimensional case, and, second, even in the finite-dimensional case, writing $\sum_j \alpha_j E_j$ often has great notational and typographical advantages over the matrix notation.

We shall make use of the fact that a not necessarily self-adjoint transformation A is isometrically diagonable (that is, that its matrix with respect to a suitable orthonormal basis is diagonal) if and only if conditions (1)–(4) of Theorem 1 hold for it. Indeed, if we have (1)–(4), then the proof of diagonability, given for self-adjoint transformations, applies; the converse we leave as an exercise for the reader.

EXERCISES

1. Suppose that A is a linear transformation on a complex inner product space. Prove that if A is Hermitian, then the linear factors of the minimal polynomial of A are distinct. Is the converse true?

2. (a) Two linear transformations A and B on a unitary space are *unitarily equivalent* if there exists a unitary transformation U such that $A = U^{-1}BU$. (The corresponding concept in the real case is called *orthogonal equivalence*.) Prove that unitary equivalence is an equivalence relation.
   (b) Are A^*A and AA^* always unitarily equivalent?
   (c) Are A and A^* always unitarily equivalent?

3. Which of the following pairs of matrices are unitarily equivalent?

(a) $\begin{pmatrix} 1 & 1 \\ 0 & 1 \end{pmatrix}$ and $\begin{pmatrix} 0 & 0 \\ 1 & 0 \end{pmatrix}$.

(b) $\begin{pmatrix} 0 & 0 & 1 \\ 0 & 0 & 0 \\ 1 & 0 & 0 \end{pmatrix}$ and $\begin{pmatrix} \frac{1}{2} & \frac{1}{2} & 0 \\ \frac{1}{2} & \frac{1}{2} & 0 \\ 0 & 0 & -1 \end{pmatrix}$.

(c) $\begin{pmatrix} 0 & 1 & 0 \\ -1 & 0 & 0 \\ 0 & 0 & -1 \end{pmatrix}$ and $\begin{pmatrix} -1 & 0 & 0 \\ 0 & i & 0 \\ 0 & 0 & i \end{pmatrix}$.

(d) $\begin{pmatrix} 0 & 1 & 0 \\ -1 & 0 & 0 \\ 0 & 0 & -1 \end{pmatrix}$ and $\begin{pmatrix} 0 & 1 & 0 \\ 1 & 0 & 0 \\ 0 & 0 & 1 \end{pmatrix}$.

4. If two linear transformations are unitarily equivalent, then they are similar, and they are congruent; if two linear transformations are either similar or congruent, then they are equivalent. Show by examples that these implication relations are the only ones that hold among these concepts.

§ 80. Normal transformations

The easiest (and at the same time the most useful) generalizations of the spectral theorem apply to complex inner product spaces (that is, unitary spaces). In order to avoid irrelevant complications, in this section we exclude the real case and concentrate attention on unitary spaces only.

We have seen that every Hermitian transformation is diagonable, and that an arbitrary transformation A may be written in the form $B + iC$, with B and C Hermitian; why isn't it true that simply by diagonalizing B and C separately we can diagonalize A? The answer is, of course, that diagonalization involves the choice of a suitable orthonormal basis, and there is no reason to expect that a basis that diagonalizes B will have the same effect on C. It is of considerable importance to know the precise class of transformations for which the spectral theorem is valid, and fortunately this class is easy to describe.

We shall call a linear transformation A *normal* if it commutes with its adjoint, $A^*A = AA^*$. (This definition makes sense, and is used, in both real and complex inner product spaces; we shall, however, continue to use techniques that are inextricably tied up with the complex case.) We point out first that A is normal if and only if its real and imaginary parts commute. Suppose, indeed, that A is normal and that $A = B + iC$ with B and C Hermitian; since $B = \frac{1}{2}(A + A^*)$ and $C = \frac{1}{2i}(A - A^*)$, it is clear that $BC = CB$. If, conversely, $BC = CB$, then the two relations $A = B + iC$ and $A^* = B - iC$ imply that A is normal. We observe that Hermitian and unitary transformations are normal.

The class of transformations possessing a spectral form in the sense of § 79 is precisely the class of normal transformations. Half of this statement is easy to prove: if $A = \sum_j \alpha_j E_j$, then $A^* = \sum_j \bar{\alpha}_j E_j$, and it takes merely a simple computation to show that $A^*A = AA^* = \sum_j |\alpha_j|^2 E_j$. To prove the converse, that is, to prove that normality implies the existence of a spectral form, we have two alternatives. We could derive this result from the spectral theorem for Hermitian transformations, using the real and imaginary parts, or we could prove that the essential lemmas of § 78, on which the proof of the Hermitian case rests, are just as valid for an arbitrary normal transformation. Because its methods are of some interest, we adopt the second procedure. We observe that the machinery needed to prove the lemmas that follow was available to us in § 78, so that we could have stated the spectral theorem for normal transformations immediately; the main reason we traveled the present course was to motivate the definition of normality.

THEOREM 1. *If A is normal, then a necessary and sufficient condition that x be a proper vector of A is that it be a proper vector of A^*; if $Ax = \lambda x$, then $A^*x = \bar{\lambda}x$.*

PROOF. We observe that the normality of A implies that

$$\| Ax \|^2 = (Ax, Ax) = (A^*Ax, x) = (AA^*x, x)$$
$$= (A^*x, A^*x) = \| A^*x \|^2.$$

Since $A - \lambda$ is normal along with A, and since $(A - \lambda)^* = A^* - \bar{\lambda}$, we obtain the relation

$$\| Ax - \lambda x \| = \| A^*x - \bar{\lambda}x \|,$$

from which the assertions of the theorem follow immediately.

THEOREM 2. *If A is normal, then proper vectors belonging to distinct proper values are orthogonal.*

PROOF. If $Ax_1 = \lambda_1 x_1$ and $Ax_2 = \lambda_2 x_2$, then

$$\lambda_1(x_1, x_2) = (Ax_1, x_2) = (x_1, A^*x_2) = \lambda_2(x_1, x_2).$$

This theorem generalizes § 78, Theorem 4; in the proof of the spectral theorem for Hermitian transformations we needed also § 78, Theorems 5 and 6. The following result takes the place of the first of these.

THEOREM 3. *If A is normal, λ is a proper value of A, and $\mathfrak{M}$ is the set of all solutions of $Ax = \lambda x$, then both $\mathfrak{M}$ and $\mathfrak{M}^{\perp}$ are invariant under A.*

PROOF. The fact that $\mathfrak{M}$ is invariant under A we have seen before; this has nothing to do with normality. To prove that $\mathfrak{M}^{\perp}$ is also invariant under A, it is sufficient to prove that $\mathfrak{M}$ is invariant under A^*. This is easy; if x is in $\mathfrak{M}$, then

$$A(A^*x) = A^*(Ax) = \lambda(A^*x),$$

so that A^*x is also in $\mathfrak{M}$.

This theorem is much weaker than its correspondent in § 78. The important thing to observe, however, is that the proof of § 78, Theorem 6, depended only on the correspondingly weakened version of Theorem 5; the only subspaces that need to be considered are the ones of the type mentioned in the preceding theorem.

This concludes the spade work; the spectral theorem for normal operators follows just as before in the Hermitian case. If in the theorems of § 79 we replace the word "self-adjoint" by "normal," delete all references to reality, and insist that the underlying inner product space be complex, the remaining parts of the statements and all the proofs remain unchanged.

It is the theory of normal transformations that is of chief interest in the study of unitary spaces. One of the most useful facts about normal transformations is that spectral conditions of the type given in § 78, Theorems 1 and 3, there shown to be necessary for the self-adjoint, positive, and isometric character of a transformation, are in the normal case also sufficient.

THEOREM 4. *A normal transformation on a finite-dimensional unitary space is* (1) *Hermitian,* (2) *positive,* (3) *strictly positive,* (4) *unitary,* (5) *invertible,* (6) *idempotent if and only if all its proper values are* (1') *real,* (2') *positive,* (3') *strictly positive,* (4') *of absolute value one,* (5') *different from zero,* (6') *equal to zero or one.*

PROOF. The fact that (1), (2), (3), and (4) imply (1'), (2'), (3'), and (4'), respectively, follows from § 78. If A is invertible and $Ax = \lambda x$, with $x \neq 0$, then $x = A^{-1}Ax = \lambda A^{-1}x$, and therefore $\lambda \neq 0$; this proves that (5) implies (5'). If A is idempotent and $Ax = \lambda x$, with $x \neq 0$, then $\lambda x = Ax = A^2 x = \lambda^2 x$, so that $(\lambda - \lambda^2)x = 0$ and therefore $\lambda = \lambda^2$; this proves that (6) implies (6'). Observe that these proofs are valid for an arbitrary inner product space (not even necessarily finite-dimensional) and that the auxiliary assumption that A is normal is also superfluous.

Suppose now that the spectral form of A is $\sum_j \alpha_j E_j$. Since $A^* = \sum_j \bar{\alpha}_j E_j$, we see that (1') implies (1). Since

$$(Ax, x) = \sum_j \alpha_j (E_j x, x) = \sum_j \alpha_j \| E_j x \|^2,$$

it follows that (2') implies (2). If $\alpha_j > 0$ for all j and if $(Ax, x) = 0$, then we must have $E_j x = 0$ for all j, and therefore $x = \sum_j E_j x = 0$; this proves that (3') implies (3). The implication from (4') to (4) follows from the relation

$$A^*A = \sum_j |\alpha_j|^2 E_j.$$

If $\alpha_j \neq 0$ for all j, we may form the linear transformation $B = \sum_j \frac{1}{\alpha_j} E_j$; since $AB = BA = 1$, it follows that (5') implies (5). Finally $A^2 = \sum_j \alpha_j^2 E_j$; from this we infer that (6') implies (6).

We observe that the implication relations (5) $\Rightarrow$ (5'), (2) $\Rightarrow$ (2'), and (3') $\Rightarrow$ (3) together fulfill a promise we made in § 72; if A is positive and invertible, then A is strictly positive.

EXERCISES

1. Give an example of a normal transformation that is neither Hermitian nor unitary.

2. (a) If A is an arbitrary linear transformation (on a finite-dimensional unitary space), and if α and β are complex numbers such that $|\alpha| = |\beta| = 1$, then $\alpha A + \beta A^*$ is normal.

(b) If $\| Ax \| = \| A^*x \|$ for all x, then A is normal.

(c) Is the sum of two normal transformations always normal?

3. If A is a normal transformation on a finite-dimensional unitary space and if $\mathfrak{M}$ is a subspace invariant under A, then the restriction of A to $\mathfrak{M}$ is also normal.

4. A linear transformation A on a finite-dimensional unitary space $\mathcal{V}$ is normal if and only if $A\mathfrak{M} \subset \mathfrak{M}$ implies $A\mathfrak{M}^\perp \subset \mathfrak{M}^\perp$ for every subspace $\mathfrak{M}$ of $\mathcal{V}$.

5. (a) If A is normal and idempotent, then it is self-adjoint.

(b) If A is normal and nilpotent, then it is zero.

(c) If A is normal and $A^3 = A^2$, then A is idempotent. Does the conclusion remain true if the assumption of normality is omitted?

(d) If A is self-adjoint and if $A^k = 1$ for some strictly positive integer k, then $A^2 = 1$.

6. If A and B are normal and if $AB = 0$, does it follow that $BA = 0$?

7. Suppose that A is a linear transformation on an n-dimensional unitary space; let $\lambda_1, \cdots, \lambda_n$ be the proper values of A (each occurring a number of times equal to its algebraic multiplicity). Prove that

$$\sum_i |\lambda_i|^2 \leqq \operatorname{tr}(A^*A),$$

and that A is normal if and only if equality holds.

8. The *numerical range* of a linear transformation A on a finite-dimensional unitary space is the set $W(A)$ of all complex numbers of the form (Ax, x), with $\| x \| = 1$.

(a) If A is normal, then $W(A)$ is convex. (This means that if ξ and η are in $W(A)$ and if $0 \leqq \alpha \leqq 1$, then $\alpha\xi + (1 - \alpha)\eta$ is also in $W(A)$.)

(b) If A is normal, then every extreme point of $W(A)$ is a proper value of A. (An extreme point is one that does not have the form $\alpha\xi + (1 - \alpha)\eta$ for any ξ and η in $W(A)$ and for any α properly between 0 and 1.)

(c) It is known that the conclusion of (a) remains true even if normality is not assumed. This fact can be phrased as follows: if A_1 and A_2 are Hermitian transformations, then the set of all points of the form $((A_1x, x), (A_2x, x))$ in the real coordinate plane (with $\| x \| = 1$) is convex. Show that the generalization of this assertion to more than two Hermitian transformations is false.

(d) Prove that the conclusion of (b) may be false for non-normal transformations.

§ 81. Orthogonal transformations

Since a unitary transformation on a unitary space is normal, the results of the preceding section include the theory of unitary transformations as a special case. Since, however, an orthogonal transformation on a real inner product space need not have any proper values, the spectral theorem, as we know it so far, gives us no information about orthogonal transformations. It is not difficult to get at the facts; the theory of complexification was made to order for this purpose.

Suppose that U is an orthogonal transformation on a finite-dimensional real inner product space $\mathcal{V}$; let U^+ be the extension of U to the complexification $\mathcal{V}^+$. Since $U^*U = 1$ (on $\mathcal{V}$), it follows that $(U^+)^*U^+ = 1$ (on $\mathcal{V}^+$), that is, that U^+ is unitary.

Let $\lambda = \alpha + i\beta$ be a complex number (α and β real), and let $\mathfrak{M}$ be the subspace consisting of all solutions of $U^+z = \lambda z$ in $\mathcal{V}^+$. (If λ is not a proper value of U^+, then $\mathfrak{M} = \Theta$.) If z is in $\mathfrak{M}$, write $z = x + iy$, with x and y in $\mathcal{V}$. The equation

$$Ux + iUy = (\alpha + i\beta)(x + iy)$$

implies (cf. § 77) that

$$Ux = \alpha x - \beta y$$

and

$$Uy = \beta x + \alpha y.$$

If we multiply the second of the last pair of equations by i and then subtract it from the first, we obtain

$$Ux - iUy = (\alpha - i\beta)(x - iy).$$

This means that $U^+\bar{z} = \bar{\lambda}\bar{z}$, where the suggestive and convenient symbol $\bar{z}$ denotes, of course, the vector $x - iy$. Since the argument (that is, the passage from $U^+z = \lambda z$ to $U^+\bar{z} = \bar{\lambda}\bar{z}$) is reversible, we have proved that the mapping $z \to \bar{z}$ is a one-to-one correspondence between $\mathfrak{M}$ and the subspace $\overline{\mathfrak{M}}$ consisting of all solutions $\bar{z}$ of $U^+\bar{z} = \bar{\lambda}\bar{z}$. The result implies, among other things, that the complex proper values of U^+ come in pairs; if λ is one of them, then so is $\bar{\lambda}$. (This remark alone we could have obtained more quickly from the fact that the coefficients of the characteristic polynomial of U^+ are real.)

We have not yet made use of the unitary character of U^+. One way we can make use of it is this. If λ is a complex (definitely not real) proper value of U^+, then $\lambda \neq \bar{\lambda}$; it follows that if $U^+z = \lambda z$, so that $U^+\bar{z} = \bar{\lambda}\bar{z}$, then z and $\bar{z}$ are orthogonal. This means that

$$0 = (x + iy, x - iy) = \|x\|^2 - \|y\|^2 + i((x, y) + (y, x)),$$

and hence that $\|x\|^2 = \|y\|^2$ and $(x, y) = -(y, x)$. Since a real inner product is symmetric ($(x, y) = (y, x)$), it follows that $(x, y) = 0$. This, in turn, implies that $\|z\|^2 = \|x\|^2 + \|y\|^2$ and hence that $\|x\| = \|y\| = \frac{1}{\sqrt{2}}\|z\|$.

If λ_1 and λ_2 are proper values of U^+ with $\lambda_1 \neq \lambda_2$ and $\lambda_1 \neq \bar{\lambda}_2$, and if $z_1 = x_1 + iy_1$ and $z_2 = x_2 + iy_2$ are corresponding proper vectors (x_1, x_2, y_1, y_2 in $\mathcal{V}$), then z_1 and z_2 are orthogonal and (since $\bar{z}_2$ is a proper vector belonging to the proper value $\bar{\lambda}_2$) z_1 and $\bar{z}_2$ are also orthogonal.

Using again the expression for the complex inner product on $\mathcal{V}^+$ in terms of the real inner product on $\mathcal{V}$, we see that

$$(x_1, x_2) + (y_1, y_2) = (x_1, y_2) - (y_1, x_2) = 0$$

and

$$(x_1, x_2) - (y_1, y_2) = (x_1, y_2) + (y_1, x_2) = 0.$$

It follows that the four vectors x_1, x_2, y_1, and y_2 are pairwise orthogonal.

The unitary transformation U^+ could have real proper values too. Since, however, we know that the proper values of U^+ have absolute value one, it follows that the only possible real proper values of U^+ are $+1$ and -1. If $U^+(x + iy) = \pm(x + iy)$, then $Ux = \pm x$ and $Uy = \pm y$, so that the proper vectors of U^+ with real proper values are obtained by putting together the proper vectors of U in an obvious manner.

We are now ready to take the final step. Given U, choose an orthonormal basis, say $\mathcal{X}_1$, in the linear manifold of solutions of $Ux = x$ (in $\mathcal{V}$), and, similarly, choose an orthonormal basis, say $\mathcal{X}_{-1}$, in the linear manifold of solutions of $Ux = -x$ (in $\mathcal{V}$). (The sets $\mathcal{X}_1$ and $\mathcal{X}_{-1}$ may be empty.) Next, for each conjugate pair of complex proper values λ and $\bar{\lambda}$ of U^+, choose an orthonormal basis $\{z_1, \cdots, z_r\}$ in the linear manifold of solutions of $U^+z = \lambda z$ (in $\mathcal{V}^+$). If $z_j = x_j + iy_j$ (with x_j and y_j in $\mathcal{V}$), let $\mathcal{X}_\lambda$ be the set $\{\sqrt{2}\,x_1, \sqrt{2}\,y_1, \cdots, \sqrt{2}\,x_r, \sqrt{2}\,y_r\}$ of vectors in $\mathcal{V}$. The results we have obtained imply that if we form the union of all the sets $\mathcal{X}_1$, $\mathcal{X}_{-1}$, and $\mathcal{X}_\lambda$, for all proper values λ of U^+, we obtain an orthonormal basis of $\mathcal{V}$. In case $\mathcal{X}_1$ has three elements, $\mathcal{X}_{-1}$ has four elements, and there are two conjugate pairs $\{\lambda_1, \bar{\lambda}_1\}$ and $\{\lambda_2, \bar{\lambda}_2\}$, then the matrix of U with respect to the basis so constructed looks like this:

$$\begin{bmatrix} 1 & & & & & & & & & & & \\ & 1 & & & & & & & & & & \\ & & 1 & & & & & & & & & \\ & & & -1 & & & & & & & & \\ & & & & -1 & & & & & & & \\ & & & & & -1 & & & & & & \\ & & & & & & -1 & & & & & \\ & & & & & & & \alpha_1 & -\beta_1 & & & \\ & & & & & & & \beta_1 & \alpha_1 & & & \\ & & & & & & & & & \alpha_2 & -\beta_2 \\ & & & & & & & & & \beta_2 & \alpha_2 \end{bmatrix}$$

(All terms not explicitly indicated are equal to zero.) In general, there is a string of $+1$'s on the main diagonal, followed by a string of -1's, and then there is a string of two-by-two boxes running down the diagonal, each box having the form $\begin{pmatrix} \alpha & -\beta \\ \beta & \alpha \end{pmatrix}$, with $\alpha^2 + \beta^2 = 1$. The fact that $\alpha^2 + \beta^2 = 1$ implies that we can find a real number θ such that $\alpha = \cos\theta$ and $\beta = \sin\theta$; it is customary to use this trigonometric representation in writing the canonical form of the matrix of an orthogonal transformation.

EXERCISES

1. Every proper value of an orthogonal transformation has absolute value 1.

2. If $A = \begin{pmatrix} 0 & 1 \\ 1 & 0 \end{pmatrix}$, how many (real) orthogonal matrices P are there with the property that $P^{-1}AP$ is diagonal?

3. State and prove a sensible analogue of the spectral theorem for normal transformations on a real inner product space.

§ 82. Functions of transformations

One of the most useful concepts in the theory of normal transformations on unitary spaces is that of a function of a transformation. If A is a normal transformation with spectral form $\sum_j \alpha_j E_j$ (for this discussion we temporarily assume that the underlying vector space is a unitary space), and if f is an arbitrary complex-valued function defined at least at the points α_j, then we define a linear transformation $f(A)$ by

$$f(A) = \sum_j f(\alpha_j) E_j.$$

Since for polynomials p (and even for rational functions) we have already seen that our earlier definition of $p(A)$ yields, if A is normal, $p(A) = \sum_j p(\alpha_j)E_j$, we see that the new notion is a generalization of the old one. The advantage of considering $f(A)$ for arbitrary functions f is for us largely notational; it introduces nothing conceptually new. Indeed, for an arbitrary f, we may write $f(\alpha_j) = \beta_j$, and then we may find a polynomial p that at the finite set of distinct complex numbers α_j takes, respectively, the values β_j. With this polynomial p we have $f(A) = p(A)$, so that the class of transformations defined by the formation of arbitrary functions is nothing essentially new; it only saves the trouble of constructing a polynomial to fit each special case. Thus for example, if, for each complex number λ, we write

$$f_\lambda(\zeta) = 0 \quad \text{whenever} \quad \zeta \neq \lambda$$

and

$$f_\lambda(\lambda) = 1,$$

then $f_\lambda(A)$ is the perpendicular projection on the subspace of solutions of $Ax = \lambda x$.

We observe that if $f(\zeta) = \dfrac{1}{\zeta}$, then (assuming of course that f is defined for all α_j, that is, that $\alpha_j \neq 0$) $f(A) = A^{-1}$, and if $f(\zeta) = \bar{\zeta}$, then $f(A) = A^*$. These statements imply that if f is an arbitrary rational function of ζ and $\bar{\zeta}$, we obtain $f(A)$ by the replacements $\zeta \to A$, $\bar{\zeta} \to A^*$, and $\dfrac{1}{\zeta} = A^{-1}$. The symbol $f(A)$ is, however, defined for much more general functions, and in the sequel we shall feel free to make use of expressions such as e^A and $\sqrt{A}$.

A particularly important function is the square root of positive transformations. We consider $f(\zeta) = \sqrt{\zeta}$, defined for all real $\zeta \geqq 0$, as the positive square root of ζ, and for every positive $A = \sum_j \alpha_j E_j$ we write

$$\sqrt{A} = \sum_j \sqrt{\alpha_j} \, E_j.$$

(Recall that $\alpha_j \geqq 0$ for all j. The discussion that follows applies to both real and complex inner product spaces.) It is clear that $\sqrt{A} \geqq 0$ and that $(\sqrt{A})^2 = A$; we should like to investigate the extent to which these properties characterize $\sqrt{A}$. At first glance it may seem hopeless to look for any uniqueness, since if we consider $B = \sum_j \pm \sqrt{\alpha_j} \, E_j$, with an arbitrary choice of sign in each place, we still have $A = B^2$. The transformation $\sqrt{A}$ that we constructed, however, was positive, and we can show that this additional property guarantees uniqueness. In other words: if $A = B^2$ and $B \geqq 0$, then $B = \sqrt{A}$. To prove this, let $B = \sum_k \beta_k F_k$ be the spectral form of B; then

$$\sum_k \beta_k{}^2 F_k = B^2 = A = \sum_j \alpha_j E_j.$$

Since the β_k are distinct and positive, so also are the $\beta_k{}^2$; the uniqueness of the spectral form of A implies that each $\beta_k{}^2$ is equal to some α_j (and vice versa), and that the corresponding E's and F's are equal. By a permutation of the indices we may therefore achieve $\beta_j{}^2 = \alpha_j$ for all j, so that $\beta_j = \sqrt{\alpha_j}$, as was to be shown.

There are several important applications of the existence of square roots for positive operators; we shall now give two of them.

First: we recall that in § 72 we mentioned three possible definitions of a positive transformation A, and adopted the weakest one, namely, that A is self-adjoint and $(Ax, x) \geqq 0$ for all x. The strongest of the three possible definitions was that we could write A in the form $A = B^2$ for some self-adjoint B. We point out that the result of this section concerning square roots implies that the (seemingly) weakest one of our conditions implies and is therefore equivalent to the strongest. (In fact, we can even obtain a unique positive square root.)

Second: in § 72 we stated also that if A and B are positive and commutative, then AB is also positive; we can now give an easy proof of this assertion. Since $\sqrt{A}$ and $\sqrt{B}$ are functions of (polynomials in) A and B respectively, the commutativity of A and B implies that $\sqrt{A}$ and $\sqrt{B}$ commute with each other; consequently

$$AB = \sqrt{A}\,\sqrt{A}\,\sqrt{B}\,\sqrt{B} = \sqrt{A}\,\sqrt{B}\,\sqrt{A}\,\sqrt{B} = (\sqrt{A}\,\sqrt{B})^2.$$

Since $\sqrt{A}$ and $\sqrt{B}$ are self-adjoint and commutative, their product is self-adjoint and therefore its square is positive.

Spectral theory also makes it quite easy to characterize the matrix (with respect to an arbitrary orthonormal coordinate system) of a positive transformation A. Since $\det A$ is the product of the proper values of A, it is clear that $A \geq 0$ implies $\det A \geq 0$. (The discussion in § 55 applies directly to complex inner product spaces only; the appropriate modification needed for the discussion of self-adjoint transformations on possibly real spaces is, however, quite easy to supply.) If we consider the defining property of positiveness expressed in terms of the matrix (α_{ij}) of A, that is, $\sum_i \sum_j \alpha_{ij}\xi_i\bar{\xi}_j \geq 0$, we observe that the last expression remains positive if we restrict the coordinates $(\xi_1, \cdots, \xi_n)$ by requiring that certain ones of them vanish. In terms of the matrix this means that if we cross out the columns numbered $j_1, \cdots, j_k$, say, and cross out also the rows bearing the *same* numbers, the remaining small matrix is still positive, and consequently so is its determinant. This fact is usually expressed by saying that the *principal minors* of the determinant of a positive matrix are positive. The converse is true. The coefficient of the j-th power of λ in the characteristic polynomial $\det (A - \lambda)$ of A is (except for sign) the sum of all principal minors of $n-j$ rows and columns. The sign is alternately plus and minus; this implies that if A has positive principal minors and is self-adjoint (so that the zeros of $\det (A - \lambda)$ are known to be real), then the proper values of A are positive. Since the self-adjoint character of a matrix is ascertainable by observing whether or not it is (Hermitian) symmetric ($\alpha_{ij} = \bar{\alpha}_{ji}$), our comments reduce the problem of finding out whether or not a matrix is positive to a finite number of elementary computations.

1. Corresponding to every unitary transformation U there is a Hermitian transformation A such that $U = e^{iA}$.

2. Discuss the theory of functions of a normal transformation on a real inner product space.

3. If $A \leqq B$ and if C is a positive transformation that commutes with both A and B, then $AC \leqq BC$.

4. A self-adjoint transformation has a unique self-adjoint cube root.

5. Find all Hermitian cube roots of the matrix

$$\begin{pmatrix} 1 & 0 & 0 \\ 0 & -1 & 0 \\ 0 & 0 & 8 \end{pmatrix}.$$

6. (a) Give an example of a linear transformation A on a finite-dimensional unitary space such that A has no square root.

(b) Prove that every Hermitian transformation on a finite-dimensional unitary space has a square root.

(c) Does every self-adjoint transformation on a finite-dimensional Euclidean space have a square root?

7. (a) Prove that if A is a positive linear transformation on a finite-dimensional inner product space, then $\rho(\sqrt{A}) = \rho(A)$.

(b) If A is a linear transformation on a finite-dimensional inner product space, is it true that $\rho(A^*A) = \rho(A)$?

8. If $A \geqq 0$ and if $(Ax, x) = 0$ for some x, then $Ax = 0$.

9. If $A \geqq 0$, then $|(Ax, y)|^2 \leqq (Ax, x)(Ay, y)$ for all x and y.

10. If the vectors $x_1, \cdots, x_k$ are linearly independent, then their Gramian is non-singular.

11. Every positive matrix is a Gramian.

12. If A and B are linear transformations on a finite-dimensional inner product space, and if $0 \leqq A \leqq B$, then $\det A \leqq \det B$. (Hint: the conclusion is trivial if $\det B = 0$; if $\det B \neq 0$, then $\sqrt{B}$ is invertible.)

13. If a linear transformation A on a finite-dimensional inner product space is strictly positive and if $A \leqq B$, then $B^{-1} \leqq A^{-1}$. (Hint: try $A = 1$ first.)

14. (a) If B is a Hermitian transformation on a finite-dimensional unitary space, then $1 + iB$ is invertible.

(b) If A is positive and invertible and if B is Hermitian, then $A + iB$ is invertible.

15. If $0 \leqq A \leqq B$, then $\sqrt{A} \leqq \sqrt{B}$. (Hint: compute

$$(\sqrt{B} + \sqrt{A} + \epsilon)(\sqrt{B} - \sqrt{A} + \epsilon),$$

and prove thereby that the second factor is invertible whenever $\epsilon > 0$.)

16. Suppose that A is a self-adjoint transformation on a finite-dimensional inner product space; write $|A| = \sqrt{A^2}$, $A_+ = \frac{1}{2}(|A| + A)$, and $A_- = \frac{1}{2}(|A| - A)$.

(a) Prove that $|A|$ is the smallest Hermitian transformation that commutes with A and for which both $A \leq |A|$ and $-A \leq |A|$. ("Smallest" refers, of course, to the ordering of Hermitian transformations.)

(b) Prove that A_+ is the smallest positive transformation that commutes with A and for which $A \leq A_+$.

(c) Prove that A_- is the smallest positive transformation that commutes with A and for which $-A \leq A_-$.

(d) Prove that if A and B are self-adjoint and commutative, then there exists a smallest self-adjoint transformation C that commutes with both A and B and for which both $A \leq C$ and $B \leq C$.

17. (a) If A and B are positive linear transformations on a finite-dimensional unitary space, and if A^2 and B^2 are unitarily equivalent, then A and B are unitarily equivalent.

(b) Is the real analogue of (a) true?

§ 83. Polar decomposition

There is another useful consequence of the theory of square roots, namely, the analogue of the polar representation $\zeta = \rho e^{i\theta}$ of a complex number.

THEOREM 1. *If A is an arbitrary linear transformation on a finite-dimensional inner product space, then there is a (uniquely determined) positive transformation P, and there is an isometry U, such that $A = UP$. If A is invertible, then U also is uniquely determined by A.*

PROOF. Although it is not logically necessary to do so, we shall first give the proof in case A is invertible; the general proof is an obvious modification of this special one, and the special proof gives greater insight into the geometric structure of the transformation A.

Since the transformation A^*A is positive, we may find its (unique) positive square root, $P = \sqrt{A^*A}$. We write $V = PA^{-1}$; since $VA = P$, the theorem will be proved if we can prove that V is an isometry, for then we may write $U = V^{-1}$. Since

$$V^* = (A^{-1})^*P^* = (A^*)^{-1}P,$$

we see that

$$V^*V = (A^*)^{-1}PPA^{-1} = (A^*)^{-1}A^*AA^{-1} = 1,$$

so that V is an isometry, and we are done.

To prove uniqueness we observe that $UP = U_0P_0$ implies $PU^* = P_0U_0^*$ and therefore

$$P^2 = PU^*UP = P_0U_0^*U_0P_0 = P_0^2.$$

Since the positive transformation $P^2 = P_0{}^2$ has only one positive square root, it follows that $P = P_0$. (In this part of the proof we did not use the invertibility of A.) If A is invertible, then so is P (since $P = U^{-1}A$), and from this we obtain (multiplying the relation $UP = U_0P_0$ on the right by $P^{-1} = P_0{}^{-1}$) that $U = U_0$.

We turn now to the general case, where we do not assume that A is invertible. We form P exactly the same way as above, so that $P^2 = A^*A$, and then we observe that for every vector x we have

$$\| Px \|^2 = (Px, Px) = (P^2x, x) = (A^*Ax, x) = \| Ax \|^2.$$

If for each vector $y = Px$ in the range $\mathfrak{R}(P)$ of P we write $Uy = Ax$, then the transformation U is length-preserving wherever it is defined. We must show that U is unambiguously determined, that is, that $Px_1 = Px_2$ implies $Ax_1 = Ax_2$. This is true since $P(x_1 - x_2) = 0$ is equivalent to $\| P(x_1 - x_2) \| = 0$ and this latter condition implies $\| A(x_1 - x_2) \| = 0$. The range of the transformation U, defined so far on the subspace $\mathfrak{R}(P)$ only, is $\mathfrak{R}(A)$. Since U is linear, $\mathfrak{R}(A)$ and $\mathfrak{R}(P)$ have the same dimension, and therefore $(\mathfrak{R}(A))^{\perp}$ and $(\mathfrak{R}(P))^{\perp}$ have the same dimension. If we define U on $(\mathfrak{R}(P))^{\perp}$ to be any linear and isometric transformation of $(\mathfrak{R}(P))^{\perp}$ onto $(\mathfrak{R}(A))^{\perp}$, then U, thereby determined on all $\mathfrak{V}$, is an isometry with the property that $UPx = Ax$ for all x. This completes the proof.

Applying the theorem just proved to A^* in place of A, and then taking adjoints, we obtain also the dual fact that every A may be written in the form $A = PU$ with an isometric U and a positive P. In contrast with the Cartesian decomposition (§ 70), we call the representation $A = UP$ a *polar decomposition* of A.

In terms of polar decompositions we obtain a new characterization of normality.

THEOREM 2. *If $A = UP$ is a polar decomposition of the linear transformation A, then a necessary and sufficient condition that A be normal is that $PU = UP$.*

PROOF. Since U is not necessarily uniquely determined by A, the statement is to be interpreted as follows: if A is normal, then P commutes with *every* U, and if P commutes with *some* U, then A is normal. Since $AA^* = UP^2U^* = UP^2U^{-1}$ and $A^*A = P^2$, it is clear that A is normal if and only if U commutes with P^2. Since, however, P^2 is a function of P and vice versa P is a function of P^2 ($P = \sqrt{P^2}$), it follows that commuting with P^2 is equivalent to commuting with P.

1. If a linear transformation on a finite-dimensional inner product space has only one polar decomposition, then it is invertible.

2. Use the functional calculus to derive the polar decomposition of a normal operator.

3. (a) If A is an arbitrary linear transformation on a finite-dimensional inner product space, then there is a partial isometry U, and there is a positive transformation P, such that $\mathfrak{R}(U) = \mathfrak{R}(P)$ and such that $A = UP$. The transformations U and P are uniquely determined by these conditions.

(b) The transformation A is normal if and only if the transformations U and P described in (a) commute with each other.

§ 84. Commutativity

The spectral theorem for self-adjoint and for normal operators and the functional calculus may also be used to solve certain problems concerning commutativity. This is a deep and extensive subject; more to illustrate some methods than for the actual results we discuss two theorems from it.

THEOREM 1. *Two self-adjoint transformations A and B on a finite-dimensional inner product space are commutative if and only if there exists a self-adjoint transformation C and there exist two real-valued functions f and g of a real variable so that $A = f(C)$ and $B = g(C)$. If such a C exists, then we may even choose C in the form $C = h(A, B)$, where h is a suitable real-valued function of two real variables.*

PROOF. The sufficiency of the condition is clear; we prove only the necessity.

Let $A = \sum_i \alpha_i E_i$ and $B = \sum_j \beta_j F_j$ be the spectral forms of A and B; since A and B commute, it follows from § 79, Theorem 3, that E_i and F_j commute. Let h be any function of two real variables such that the numbers $h(\alpha_i, \beta_j) = \gamma_{ij}$ are all distinct, and write

$$C = h(A, B) = \sum_i \sum_j h(\alpha_i, \beta_j) E_i F_j.$$

(It is clear that h may even be chosen as a polynomial, and the same is true of the functions f and g we are about to describe.) Let f and g be such that $f(\gamma_{ij}) = \alpha_i$ and $g(\gamma_{ij}) = \beta_j$ for all i and j. It follows that $f(C) = A$ and $g(C) = B$, and everything is proved.

THEOREM 2. *If A is a normal transformation on a finite-dimensional unitary space and if B is an arbitrary transformation that commutes with A, then B commutes with A^*.*

PROOF. Let $A = \sum_i \alpha_i E_i$ be the spectral form of A; then $A^* = \sum_i \bar{\alpha}_i E_i$. Let f be such a function (polynomial) of a complex variable that $f(\alpha_i) = \bar{\alpha}_i$ for all i. Since $A^* = f(A)$, the conclusion follows.

EXERCISES

1. (a) Prove the following generalization of Theorem 2: if A_1 and A_2 are normal transformations (on a finite-dimensional unitary space) and if $A_1 B = B A_2$, then $A_1^* B = B A_2^*$.

(b) Theorem 2 asserts that the relation of commutativity is sometimes transitive: if A^* commutes with A and if A commutes with B, then A^* commutes with B. Does this formulation remain true if A^* is replaced by an arbitrary transformation C?

2. (a) If A commutes with A^*A, does it follow that A is normal?

(b) If A^*A commutes with AA^*, does it follow that A is normal?

3. (a) A linear transformation A is normal if and only if there exists a polynomial p such that $A^* = p(A)$.

(b) If A is normal and commutes with B, then A commutes with B^*.

(c) If A and B are normal and commutative, then AB is normal.

4. If A and B are normal and similar, then they are unitarily equivalent.

5. (a) If A is Hermitian, if every proper value of A has multiplicity 1, and if $AB = BA$, then there exists a polynomial p such that $B = p(A)$.

(b) If A is Hermitian, then a necessary and sufficient condition that there exist a polynomial p such that $B = p(A)$ is that B commute with every linear transformation that commutes with A.

6. Show that a commutative set of normal transformations on a finite-dimensional unitary space can be simultaneously diagonalized.

§ 85. Self-adjoint transformations of rank one

We have already seen (§ 51, Theorem 2) that every linear transformation A of rank ρ is the sum of ρ linear transformations of rank one. It is easy to see (using the spectral theorem) that if A is self-adjoint, or positive, then the summands may also be taken self-adjoint, or positive, respectively. We know (§ 51, Theorem 1) what the matrix of transformation of rank one has to be; what more can we say if the transformation is self-adjoint or positive?

THEOREM 1. *If A has rank one and is self-adjoint (or positive), then in every orthonormal coordinate system the matrix (α_{ij}) of A is given by $\alpha_{ij} = \kappa \beta_i \bar{\beta}_j$ with a real κ (or by $\alpha_{ij} = \gamma_i \bar{\gamma}_j$). If, conversely, $[A]$ has this form in some orthonormal coordinate system, then A has rank one and is self-adjoint (or positive).*

PROOF. We know that the matrix (α_{ij}) of a transformation A of rank one, in any orthonormal coordinate system $\mathfrak{X} = \{x_1, \cdots, x_n\}$, is given by $\alpha_{ij} = \beta_i \gamma_j$. If A is self-adjoint, we must also have $\alpha_{ij} = \bar{\alpha}_{ji}$, whence $\beta_i \gamma_j = \bar{\beta_j \gamma_i}$. If $\beta_i = 0$ and $\gamma_i \neq 0$ for some i, then $\bar{\beta}_j = \beta_i \gamma_j / \bar{\gamma}_i = 0$ for all j, whence $A = 0$. Since we assumed that the rank of A is one (and not zero), this is impossible. Similarly $\beta_i \neq 0$ and $\gamma_i = 0$ is impossible; that is, we can find an i for which $\beta_i \gamma_i \neq 0$. Using this i, we have

$$\bar{\beta}_j = (\beta_i / \bar{\gamma}_i) \gamma_j = \kappa \gamma_j$$

with some non-zero constant κ, independent of j. Since the diagonal elements $\alpha_{jj} = (Ax_j, x_j) = \beta_j \gamma_j$ of a self-adjoint matrix are real, we can even conclude that $\alpha_{ij} = \kappa \bar{\beta}_i \beta_j$ with a real κ.

If, moreover, A is positive, then we even know that $\kappa \beta_j \bar{\beta}_j = \alpha_{jj} = (Ax_j, x_j)$ is positive, and therefore so is κ. In this case we write $\lambda = \sqrt{\kappa}$; the relation $\kappa \beta_i \bar{\beta}_j = (\lambda \beta_i)(\lambda \bar{\beta}_j)$ shows that α_{ij} is given by $\alpha_{ij} = \gamma_i \bar{\gamma}_j$.

It is easy to see that these necessary conditions are also sufficient. If $\alpha_{ij} = \kappa \bar{\beta}_i \beta_j$ with a real κ, then A is self-adjoint. If $\alpha_{ij} = \gamma_i \bar{\gamma}_j$, and $x = \sum_i \xi_i x_i$, then

$$(Ax, x) = \sum_i \sum_j \alpha_{ij} \bar{\xi}_i \xi_j = \sum_i \sum_j \gamma_i \bar{\gamma}_j \bar{\xi}_i \xi_j$$

$$= \left(\sum_i \gamma_i \bar{\xi}_i\right)\overline{\left(\sum_j \gamma_j \bar{\xi}_j\right)} = \left|\sum_i \gamma_i \bar{\xi}_i\right|^2 \geq 0,$$

so that A is positive.

As a consequence of Theorem 1 it is very easy to prove a remarkable theorem on positive matrices.

THEOREM 2. *If A and B are positive linear transformations whose matrices in some orthonormal coordinate system are (α_{ij}) and (β_{ij}) respectively, then the linear transformation C, whose matrix (γ_{ij}) in the same coordinate system is given by $\gamma_{ij} = \alpha_{ij}\beta_{ij}$ for all i and j, is also positive.*

PROOF. Since we may write both A and B as sums of positive transformations of rank one, so that

$$\alpha_{ij} = \sum_p \alpha_i^p \bar{\alpha}_j^p$$

and

$$\beta_{ij} = \sum_q \beta_i^q \bar{\beta}_j^q,$$

it follows that

$$\gamma_{ij} = \sum_p \sum_q \alpha_i^p \beta_i^q \overline{(\alpha_j^p \beta_j^q)}.$$

(The superscripts here are not exponents.) Since a sum of positive matrices is positive, it will be sufficient to prove that, for each fixed p and q, the matrix $((\alpha_i^p \beta_i^q)(\overline{\alpha_j^p \beta_j^q}))$ is positive, and this follows from Theorem 1.

The proof shows, by the way, that Theorem 2 remains valid if we replace "positive" by "self-adjoint" in both hypothesis and conclusion; in most applications, however, it is only the actually stated version that is useful. The matrix (γ_{ij}) described in Theorem 2 is called the *Hadamard product* of (α_{ij}) and (β_{ij}).

1. Suppose that $\mathfrak{U}$ and $\mathfrak{V}$ are finite-dimensional inner product spaces (both real or both complex).

(a) There is a unique inner product on the vector space of all bilinear forms on $\mathfrak{U} \oplus \mathfrak{V}$ such that if $w_1(x, y) = (x, x_1)(y, y_1)$ and $w_2(x, y) = (x, x_2)(y, y_2)$, then $(w_1, w_2) = (x_2, x_1)(y_2, y_1)$.

(b) There is a unique inner product on the tensor product $\mathfrak{U} \otimes \mathfrak{V}$ such that if $z_1 = x_1 \otimes y_1$ and $z_2 = x_2 \otimes y_2$, then $(z_1, z_2) = (x_1, x_2)(y_1, y_2)$.

(c) If $\{x_i\}$ and $\{y_p\}$ are orthonormal bases in $\mathfrak{U}$ and $\mathfrak{V}$, respectively, then the vectors $x_i \otimes y_p$ form an orthonormal basis in $\mathfrak{U} \otimes \mathfrak{V}$.

2. Is the tensor product of two Hermitian transformations necessarily Hermitian? What about unitary transformations? What about normal transformations?

CHAPTER IV

ANALYSIS

Essentially the only way in which we exploited, so far, the existence of an inner product in an inner product space was to introduce the notion of a normal transformation together with certain important special cases of it. A much more obvious circle of ideas is the study of the convergence problems that arise in an inner product space.

Let us see what we might mean by the assertion that a sequence (x_n) of vectors in $\mathcal{V}$ converges to a vector x in $\mathcal{V}$. There are two possibilities that suggest themselves:

(i) $$\| x_n - x \| \rightarrow 0 \text{ as } n \rightarrow \infty;$$

(ii) $$(x_n - x, y) \rightarrow 0 \text{ as } n \rightarrow \infty, \text{ for each fixed } y \text{ in } \mathcal{V}.$$

If (i) is true, then we have, for every y,

$$| (x_n - x, y) | \leqq \| x_n - x \| \cdot \| y \| \rightarrow 0,$$

so that (ii) is true. In a finite-dimensional space the converse implication is valid: (ii) $\Rightarrow$ (i). To prove this, let $\{z_1, \cdots, z_N\}$ be an orthonormal basis in $\mathcal{V}$. (Often in this chapter we shall write N for the dimension of a finite-dimensional vector space, in order to reserve n for the dummy variable in limiting processes.) If we assume (ii), then $(x_n - x, z_i) \rightarrow 0$ for each $i = 1, \cdots, N$. Since (§ 63, Theorem 2)

$$\| x_n - x \|^2 = \sum_i | (x_n - x, z_i) |^2,$$

it follows that $\| x_n - x \| \rightarrow 0$, as was to be proved.

Concerning the convergence of vectors (in either of the two equivalent senses) we shall use without proof the following facts. (All these facts are easy consequences of our definitions and of the properties of convergence in the usual domain of complex numbers; we assume that the reader has a modicum of familiarity with these notions.) The expression $\alpha x + \beta y$ defines a continuous function of all its arguments simultaneously; that is, if (α_n) and (β_n) are sequences of numbers and (x_n) and (y_n) are sequences of vectors, then $\alpha_n \to \alpha$, $\beta_n \to \beta$, $x_n \to x$, and $y_n \to y$ imply that $\alpha_n x_n + \beta_n y_n \to \alpha x + \beta y$. If $\{z_i\}$ is an orthonormal basis in $\mathcal{U}$, and if $x_n = \sum_i \alpha_{in} z_i$ and $x = \sum_i \alpha_i z_i$, then a necessary and sufficient condition that $x_n \to x$ is that $\alpha_{in} \to \alpha_i$ (as $n \to \infty$) for each $i = 1, \cdots, N$. (Thus the notion of convergence here defined coincides with the usual one in N-dimensional real or complex coordinate space.) Finally, we shall assume as known the fact that a finite-dimensional inner product space with the metric defined by the norm is complete; that is, if (x_n) is a sequence of vectors for which $\| x_n - x_m \| \to 0$, as $n, m \to \infty$, then there is a (unique) vector x such that $x_n \to x$ as $n \to \infty$.

§ 87. Norm

The metric properties of vectors have certain important implications for the metric properties of linear transformations, which we now begin to study.

DEFINITION. A linear transformation A on an inner product space $\mathcal{U}$ is *bounded* if there exists a constant K such that $\| Ax \| \leq K \| x \|$ for every vector x in $\mathcal{U}$. The greatest lower bound of all constants K with this property is called the *norm* (or *bound*) of A and is denoted by $\| A \|$.

Clearly if A is bounded, then $\| Ax \| \leq \| A \| \cdot \| x \|$ for all x. For examples we may consider the cases where A is a (non-zero) perpendicular projection or an isometry; § 75, Theorem 1, and the theorem of § 73, respectively, imply that in both cases $\| A \| = 1$. Considerations of the vectors defined by $x_n(t) = t^n$ in $\mathcal{O}$ shows that the differentiation transformation is not bounded.

Because in the sequel we shall have occasion to consider quite a few upper and lower bounds similar to $\| A \|$, we introduce a convenient notation. If P is any possible property of real numbers t, we shall denote the set of all real numbers t possessing the property P by the symbol $\{t: P\}$, and we shall denote greatest lower bound and least upper bound by *inf* (for infimum) and *sup* (for supremum) respectively. In this notation we have, for example,

$$\| A \| = \inf \{K : \| Ax \| \leq K \| x \| \text{ for all } x\}.$$

The notion of boundedness is closely connected with the notion of continuity. If A is bounded and if ϵ is any positive number, by writing

$\delta = \dfrac{\epsilon}{\| A \|}$ we may make sure that $\| x - y \| < \delta$ implies that

$$\| Ax - Ay \| = \| A(x - y) \| \leqq \| A \| \cdot \| x - y \| < \epsilon;$$

in other words boundedness implies (uniform) continuity. (In this proof we tacitly assumed that $\| A \| \neq 0$; the other case is trivial.) In view of this fact the following result is a welcome one.

THEOREM. *Every linear transformation on a finite-dimensional inner product space is bounded.*

PROOF. Suppose that A is a linear transformation on $\mathcal{U}$; let $\{x_1, \cdots, x_N\}$ be an orthonormal basis in $\mathcal{U}$ and write

$$K_0 = \max \{\| Ax_1 \|, \cdots, \| Ax_N \|\}.$$

Since an arbitrary vector x may be written in the form $x = \sum_i (x, x_i)x_i$, we obtain, applying the Schwarz inequality and remembering that $\| x_i \| = 1$,

$$\begin{aligned}
\| Ax \| &= \| A(\textstyle\sum_i (x, x_i)x_i) \| \\
&= \| \textstyle\sum_i (x, x_i)Ax_i \| \leqq \textstyle\sum_i |(x, x_i)| \cdot \| Ax_i \| \\
&\leqq \textstyle\sum_i \| x \| \cdot \| x_i \| \cdot \| Ax_i \| \leqq K_0 \textstyle\sum_i \| x \| \\
&= NK_0 \| x \|.
\end{aligned}$$

In other words, $K = NK_0$ is a bound of A, and the proof is complete.

It is no accident that the dimension N of $\mathcal{U}$ enters into our evaluation; we have already seen that the theorem is not true in infinite-dimensional spaces.

EXERCISES

1. (a) Prove that the inner product is a continuous function (and therefore so also is the norm); that is, if $x_n \to x$ and $y_n \to y$, then $(x_n, y_n) \to (x, y)$.
   (b) Is every linear functional continuous? How about multilinear forms?

2. A linear transformation A on an inner product space is said to be bounded from below if there exists a (strictly) positive constant K such that $\| Ax \| \geqq K \| x \|$ for every x. Prove that (on a finite-dimensional space) A is bounded from below if and only if it is invertible.

3. If a linear transformation on an inner product space (not necessarily finite-dimensional) is continuous at one point, then it is bounded (and consequently continuous over the whole space).

4. For each positive integer n construct a projection E_n (not a perpendicular projection) such that $\| E_n \| \geqq n$.

5. (a) If U is a partial isometry other than 0, then $\| U \| = 1$.
   (b) If U is an isometry, then $\| UA \| = \| AU \| = \| A \|$ for every linear transformation A.

6. If E and F are perpendicular projections, with ranges $\mathfrak{M}$ and $\mathfrak{N}$ respectively, and if $\| E - F \| < 1$, then dim $\mathfrak{M}$ = dim $\mathfrak{N}$.

7. (a) If A is normal, then $\| A^n \| = \| A \|^n$ for every positive integer n.

(b) If A is a linear transformation on a 2-dimensional unitary space and if $\| A^2 \| = \| A \|^2$, then A is normal.

(c) Is the conclusion of (b) true for transformations on a 3-dimensional space?

§ 88. Expressions for the norm

To facilitate working with the norm of a transformation, we consider the following four expressions:

$$p = \sup \{ \| Ax \| / \| x \| : x \neq 0 \},$$

$$q = \sup \{ \| Ax \| : \| x \| = 1 \},$$

$$r = \sup \{ |(Ax, y)| / \| x \| \cdot \| y \| : x \neq 0, y \neq 0 \},$$

$$s = \sup \{ |(Ax, y)| : \| x \| = \| y \| = 1 \}.$$

In accordance with our definition of the brace notation, the expression $\{ \| Ax \| : \| x \| = 1 \}$, for example, means the set of all real numbers of the form $\| Ax \|$, considered for all x's for which $\| x \| = 1$.

Since $\| Ax \| \leq K \| x \|$ is trivially true with any K if $x = 0$, the definition of supremum implies that $p = \| A \|$; we shall prove that, in fact, $p = q = r = s = \| A \|$. Since the supremum in the expression for q is extended over a subset of the corresponding set for p (that is, if $\| x \| = 1$, then $\| Ax \| / \| x \| = \| Ax \|$), we see that $q \leq p$; a similar argument shows that $s \leq r$.

For any $x \neq 0$ we consider $y = \dfrac{x}{\| x \|}$ (so that $\| y \| = 1$); we have $\| Ax \| / \| x \| = \| Ay \|$. In other words, every number of the set whose supremum is p occurs also in the corresponding set for q; it follows that $p \leq q$, and consequently that $p = q = \| A \|$.

Similarly if $x \neq 0$ and $y \neq 0$, we consider $x' = x / \| x \|$ and $y' = y / \| y \|$; we have

$$|(Ax, y)| / \| x \| \cdot \| y \| = |(Ax', y')|,$$

and hence, by the argument just used, $r \leq s$, so that $r = s$.

To consolidate our position, we note that so far we have proved that

$$p = q = \| A \| \quad \text{and} \quad r = s.$$

Since

$$\frac{|(Ax, y)|}{\| x \| \cdot \| y \|} \leq \frac{\| Ax \| \cdot \| y \|}{\| x \| \cdot \| y \|} = \frac{\| Ax \|}{\| x \|},$$

it follows that $r \leq p$; we shall complete the proof by showing that $p \leq r$.

For this purpose we consider any vector x for which $Ax \neq 0$ (so that $x \neq 0$); for such an x we write $y = Ax$ and we have

$$\| Ax \|/\| x \| = |(Ax, y)|/\| x \| \cdot \| y \|.$$

In other words, we proved that every number that occurs in the set defining p, and is different from zero, occurs also in the set of which r is the supremum; this clearly implies the desired result.

The numerical function of a transformation A given by $\| A \|$ satisfies the following four conditions:

(1) $$\| A + B \| \leq \| A \| + \| B \|,$$

(2) $$\| AB \| \leq \| A \| \cdot \| B \|,$$

(3) $$\| \alpha A \| = |\alpha| \cdot \| A \|,$$

(4) $$\| A^* \| = \| A \|.$$

The proof of the first three of these is immediate from the definition of the norm of a transformation; for the proof of (4) we use the equation $\| A \| = r$, as follows. Since

$$|(Ax, y)| = |(x, A^*y)| \leq \| x \| \cdot \| A^*y \|$$

$$\leq \| A^* \| \cdot \| x \| \cdot \| y \|,$$

we see that $\| A \| \leq \| A^* \|$; replacing A by A^* and A^* by $A^{**} = A$, we obtain the reverse inequality.

EXERCISES

1. If B is invertible, then $\| AB \| \geq \| A \|/\| B^{-1} \|$ for every A.

2. Is it true for every linear transformation A that $\| A^*A \| = \| AA^* \|$?

3. (a) If A is Hermitian and if $\alpha \geq 0$, then a necessary and sufficient condition that $\| A \| \leq \alpha$ is that $-\alpha \leq A \leq \alpha$.
   (b) If A is Hermitian, if $\alpha \leq A \leq \beta$, and if p is a polynomial such that $p(t) \geq 0$ whenever $\alpha \leq t \leq \beta$, then $p(A) \geq 0$.
   (c) If A is Hermitian, if $\alpha \leq A \leq \beta$, and if p is a polynomial such that $p(t) \neq 0$ whenever $\alpha \leq t \leq \beta$, then $p(A)$ is invertible.

§ 89. Bounds of a self-adjoint transformation

As usual we can say a little more about the special case of self-adjoint transformations than in the general case. We consider, for any self-adjoint transformation A, the sets of real numbers

$$\Phi = \{(Ax, x)/\| x \|^2 : x \neq 0\}$$

and

$$\Psi = \{(Ax, x) : \| x \| = 1\}.$$

It is clear that $\Psi \subset \Phi$. If, for every $x \neq 0$, we write $y = x/\|x\|$, then $\|y\| = 1$ and $(Ax, x)/\|x\|^2 = (Ay, y)$, so that every number in Φ occurs also in Ψ and consequently $\Phi = \Psi$. We write

$$\alpha = \inf \Phi = \inf \Psi,$$

$$\beta = \sup \Phi = \sup \Psi,$$

and we say that α is the *lower bound* and β is the *upper bound* of the self-adjoint transformation A. If we recall the definition of a positive transformation, we see that α is the greatest real number for which $A - \alpha \geq 0$ and β is the least real number for which $\beta - A \geq 0$. Concerning these numbers we assert that

$$\gamma = \max \{|\alpha|, |\beta|\} = \|A\|.$$

Half the proof is easy. Since

$$|(Ax, x)| \leq \|Ax\| \cdot \|x\| \leq \|A\| \cdot \|x\|^2,$$

it is clear that both $|\alpha|$ and $|\beta|$ are dominated by $\|A\|$. To prove the reverse inequality, we observe that the positive character of the two linear transformations $\gamma - A$ and $\gamma + A$ implies that both

$$(\gamma + A)^*(\gamma - A)(\gamma + A) = (\gamma + A)(\gamma - A)(\gamma + A)$$

and

$$(\gamma - A)^*(\gamma + A)(\gamma - A) = (\gamma - A)(\gamma + A)(\gamma - A)$$

are positive, and, therefore, so also is their sum $2\gamma(\gamma^2 - A^2)$. Since $\gamma = 0$ implies $\|A\| = 0$, the assertion is trivial in this case; in any other case we may divide by 2 and obtain the result that $\gamma^2 - A^2 \geq 0$. In other words,

$$\gamma^2\|x\|^2 = \gamma^2(x, x) \geq (A^2x, x) = \|Ax\|^2,$$

whence $\gamma \geq \|A\|$, and the proof is complete.

We call the reader's attention to the fact that the computation in the main body of this proof could have been avoided entirely. Since both $\gamma - A$ and $\gamma + A$ are positive, and since they commute, we may conclude immediately (§ 82) that their product $\gamma^2 - A^2$ is positive. We presented the roundabout method in accordance with the principle that, with an eye to the generalizations of the theory, one should avoid using the spectral theorem whenever possible. Our proof of the fact that the positiveness and commutativity of A and B imply the positiveness of AB was based on the existence of square roots for positive transformations. This fact, to be sure, can be obtained by so-called "elementary" methods, that is, methods not using the spectral theorem, but even the simplest elementary proof involves complications that are purely technical and, for our purposes, not particularly useful.

§ 90. Minimax principle

A very elegant and useful fact concerning self-adjoint transformations is the following *minimax principle*.

THEOREM. *Let A be a self-adjoint transformation on an n-dimensional inner product space $\mathcal{V}$, and let $\lambda_1, \cdots, \lambda_n$ be the (not necessarily distinct) proper values of A, with the notation so chosen that $\lambda_1 \geqq \lambda_2 \geqq \cdots \geqq \lambda_n$. If, for each subspace $\mathfrak{M}$ of $\mathcal{V}$,*

$$\mu(\mathfrak{M}) = \sup \{(Ax, x): x \text{ in } \mathfrak{M}, \| x \| = 1\},$$

and if, for $k = 1, \cdots, n$,

$$\mu_k = \inf \{\mu(\mathfrak{M}): \dim \mathfrak{M} = n - k + 1\},$$

then $\mu_k = \lambda_k$ for $k = 1, \cdots, n$.

PROOF. Let $\{x_1, \cdots, x_n\}$ be an orthonormal basis in $\mathcal{V}$ for which $Ax_i = \lambda_i x_i, i = 1, \cdots, n$ (§ 79); let $\mathfrak{M}_k$ be the subspace spanned by $x_1, \cdots, x_k$, for $k = 1, \cdots, n$. Since the dimension of $\mathfrak{M}_k$ is k, the subspace $\mathfrak{M}_k$ cannot be disjoint from any $(n - k + 1)$-dimensional subspace $\mathfrak{M}$ in $\mathcal{V}$; if $\mathfrak{M}$ is any such subspace, we may find a vector x belonging to both $\mathfrak{M}_k$ and $\mathfrak{M}$ and such that $\| x \| = 1$. For this $x = \sum_{i=1}^{k} \xi_i x_i$ we have

$$(Ax, x) = \sum_{i=1}^{k} \lambda_i |\xi_i|^2 \geqq \lambda_k \sum_{i=1}^{k} |\xi_i|^2$$
$$= \lambda_k \| x \|^2 = \lambda_k,$$

so that $\mu(\mathfrak{M}) \geqq \lambda_k$.

If, on the other hand, we consider the particular $(n - k + 1)$-dimensional subspace $\mathfrak{M}_0$ spanned by $x_k, x_{k+1}, \cdots, x_n$, then, for each $x = \sum_{i=k}^{n} \xi_i x_i$ in this subspace, we have (assuming $\| x \| = 1$)

$$(Ax, x) = \sum_{i=k}^{n} \lambda_i |\xi_i|^2 \leqq \lambda_k \sum_{i=k}^{n} |\xi_i|^2$$
$$= \lambda_k \| x \|^2 = \lambda_k,$$

so that $\mu(\mathfrak{M}_0) \leqq \lambda_k$.

In other words, as $\mathfrak{M}$ runs over all $(n - k + 1)$-dimensional subspaces, $\mu(\mathfrak{M})$ is always $\geqq \lambda_k$, and is at least once $\leqq \lambda_k$; this shows that $\mu_k = \lambda_k$, as was to be proved.

In particular for $k = 1$ we see (using § 89) that if A is self-adjoint, then $\| A \|$ is equal to the maximum of the absolute values of the proper values of A.

1. If λ is a proper value of a linear transformation A on a finite-dimensional inner product space, then $|\lambda| \leq \| A \|$.

2. If A and B are linear transformations on a finite-dimensional unitary space, and if $C = AB - BA$, then $\| 1 - C \| \geq 1$. (Hint: consider the proper values of C.)

3. If A and B are linear transformations on a finite-dimensional unitary space, if $C = AB - BA$, and if C commutes with A, then C is not invertible. (Hint: if C is invertible, then $2\| B \| \cdot \| A \| \cdot \|A^{k-1}\| \geq k \| A^{k-1}\|/\| C^{-1}\|$.)

4. (a) If A is a normal linear transformation on a finite-dimensional unitary space, then $\| A \|$ is equal to the maximum of the absolute values of the proper values of A.

(b) Does the conclusion of (a) remain true if the hypothesis of normality is omitted?

5. The *spectral radius* of a linear transformation A on a finite-dimensional unitary space, denoted by $r(A)$, is the maximum of the absolute values of the proper values of A.

(a) If $f(\lambda) = ((1 - \lambda A)^{-1}x, y)$, then f is an analytic function of λ in the region determined by $|\lambda| < \dfrac{1}{r(A)}$ (for each fixed x and y).

(b) There exists a constant K such that $|\lambda|^n \| A^n \| \leq K$ whenever $|\lambda| < \dfrac{1}{r(A)}$ and $n = 0, 1, 2, \cdots$. (Hint: for each x and y there exists a constant K such that $|\lambda^n(A^n x, y)| \leq K$ for all n.)

(c) $\lim \sup_n \| A^n \|^{1/n} \leq r(A)$.

(d) $(r(A))^n \leq r(A^n)$, $n = 0, 1, 2, \cdots$.

(e) $r(A) = \lim_n \| A^n \|^{1/n}$.

6. If A is a linear transformation on a finite-dimensional unitary space, then a necessary and sufficient condition that $r(A) = \| A \|$ is that $\| A^n \| = \| A \|^n$ for $n = 0, 1, 2, \cdots$.

7. (a) If A is a positive linear transformation on a finite-dimensional inner product space, and if AB is self-adjoint, then

$$|(ABx, x)| \leq \| B \| \cdot (Ax, x)$$

for every vector x.

(b) Does the conclusion of (a) remain true if $\| B \|$ is replaced by $r(B)$?

§ 91. Convergence of linear transformations

We return now to the consideration of convergence problems. There are three obvious senses in which we may try to define the convergence of a sequence (A_n) of linear transformations to a fixed linear transformation A.

(i) $$\| A_n - A \| \to 0 \text{ as } n \to \infty.$$

(ii) $\| A_n x - A x \| \to 0$ as $n \to \infty$ for each fixed x.

(iii) $|(A_n x, y) - (A x, y)| \to 0$ as $n \to \infty$ for each fixed x and y.

If (i) is true, then, for every x,

$$\| A_n x - A x \| = \| (A_n - A)x \| \leq \| A_n - A \| \cdot \| x \| \to 0,$$

so that (i) $\Rightarrow$ (ii). We have already seen (§ 86) that (ii) $\Rightarrow$ (iii) and that in finite-dimensional spaces (iii) $\Rightarrow$ (ii). It is even true that in finite-dimensional spaces (ii) $\Rightarrow$ (i), so that all three conditions are equivalent. To prove this, let $\{x_1, \cdots, x_N\}$ be an orthonormal basis in $\mathcal{U}$. If we suppose that (ii) holds, then, for each $\epsilon > 0$, we may find an $n_0 = n_0(\epsilon)$ such that $\| A_n x_i - A x_i \| < \epsilon$ for $n \geq n_0$ and for $i = 1, \cdots, N$. It follows that for an arbitrary $x = \sum_i (x, x_i) x_i$ we have

$$\| (A_n - A)x \| = \| \sum_i (x, x_i)(A_n - A)x_i \|$$

$$\leq \sum_i \| x \| \cdot \| (A_n - A)x_i \| \leq \epsilon N \| x \|,$$

and this implies (i).

It is also easy to prove that if the norm is used to define a distance for transformations, then the resulting metric space is complete, that is, if $\| A_n - A_m \| \to 0$ as $n, m \to \infty$, then there is an A such that $\| A_n - A \| \to 0$. The proof of this fact is reduced to the corresponding fact for vectors. If $\| A_n - A_m \| \to 0$, then $\| A_n x - A_m x \| \to 0$ for each x, so that we may find a vector corresponding to x, which we may denote by $A x$, say, such that $\| A_n x - A x \| \to 0$. It is clear that the correspondence from x to $A x$ is given by a linear transformation A; the implication relation (ii) $\Rightarrow$ (i) proved above completes the proof.

Now that we know what convergence means for linear transformations, it behooves us to examine some simple functions of these transformations in order to verify their continuity. We assert that $\| A \|$, $\| A x \|$, $(A x, y)$, $A x$, $A + B$, αA, $A B$, and A^* all define continuous functions of all their arguments simultaneously. (Observe that the first three are numerical-valued functions, the next is vector-valued, and the last four are transformation-valued.) The proofs of these statements are all quite easy, and similar to each other; to illustrate the ideas we discuss $\| A \|$, $A x$, and A^*.

(1) If $A_n \to A$, that is, $\| A_n - A \| \to 0$, then, since the relations

$$\| A_n \| \leq \| A_n - A \| + \| A \|,$$

and

$$\| A \| \leq \| A - A_n \| + \| A_n \|,$$

imply that

$$| \| A_n \| - \| A \| | \leq \| A_n - A \|,$$

we see that $\| A_n \| \to \| A \|$.

(2) If $A_n \to A$ and $x_n \to x$, then

$$\| A_n x_n - A x \| \leqq \| A_n x_n - A x_n \| + \| A x_n - A x \| \to 0,$$

so that $A_n x_n \to A x$.

(3) If $A_n \to A$, then, for each x and y,

$$(A_n^* x, y) = (x, A_n y) = \overline{(A_n y, x)} \to \overline{(A y, x)}$$

$$= \overline{(y, A^* x)} = (A^* x, y),$$

whence $A_n^* \to A^*$.

1. A sequence (A_n) of linear transformations converges to a linear transformation A if and only if, for every coordinate system, each entry in the matrix of A_n converges, as $n \to \infty$, to the corresponding entry in the matrix of A.

2. For every linear transformation A there exists a sequence (A_n) of invertible linear transformations such that $A_n \to A$.

3. If E and F are perpendicular projections, then $(EFE)^n$ converges, as $n \to \infty$, to the projection whose range is the intersection of the ranges of E and F.

4. If A is a linear transformation on a finite-dimensional unitary space, then a necessary and sufficient condition that $A^n \to 0$ is that all the proper values of A be (strictly) less than 1 in absolute value.

5. Prove that if A is the n-by-n matrix

$$\begin{bmatrix} 0 & 1 & 0 & \cdots & 0 \\ 0 & 0 & 1 & \cdots & 0 \\ \cdot & \cdot & \cdot & \cdots & \cdot \\ 0 & 0 & 0 & \cdots & 1 \\ \frac{1}{n} & \frac{1}{n} & \frac{1}{n} & \cdots & \frac{1}{n} \end{bmatrix}$$

then A^k converges, as $k \to \infty$, to a projection whose range is one-dimensional; find the range.

6. Prove that *det* and *tr* are continuous.

§ 92. Ergodic theorem

The routine work is out of the way; we go on to illustrate the general theory by considering some very special but quite important convergence problems.

THEOREM. *If U is an isometry on a finite-dimensional inner product space, and if $\mathfrak{M}$ is the subspace of all solutions of $Ux = x$, then the sequence defined by*

$$V_n = \frac{1}{n}(1 + U + \cdots + U^{n-1})$$

converges as $n \to \infty$ to the perpendicular projection $E = P_{\mathfrak{M}}$.

PROOF. Let $\mathfrak{N}$ be the range of the linear transformation $1 - U$. If $x = y - Uy$ is in $\mathfrak{N}$, then

$$V_n x = \frac{1}{n}(y - Uy + Uy - U^2y + \cdots + U^{n-1}y - U^n y)$$

$$= \frac{1}{n}(y - U^n y),$$

so that

$$\| V_n x \| = \frac{1}{n} \| y - U^n y \| \leq \frac{1}{n}(\| y \| + \| U^n y \|)$$

$$= \frac{2}{n} \| y \|.$$

This implies that $V_n x$ converges to zero when x is in $\mathfrak{N}$.

On the other hand, if x is in $\mathfrak{M}$, that is, $Ux = x$, then $V_n x = x$, so that in this case $V_n x$ certainly converges to x.

We shall complete the proof by showing that $\mathfrak{N}^{\perp} = \mathfrak{M}$. (This will imply that every vector is a sum of two vectors for which (V_n) converges, so that (V_n) converges everywhere. What we have already proved about the limit of (V_n) in $\mathfrak{M}$ and in $\mathfrak{N}$ shows that $(V_n x)$ always converges to the projection of x in $\mathfrak{M}$.) To show that $\mathfrak{N}^{\perp} = \mathfrak{M}$, we observe that x is in the orthogonal complement of $\mathfrak{N}$ if and only if $(x, y - Uy) = 0$ for all y. This in turn implies that

$$0 = (x, y - Uy) = (x, y) - (x, Uy) = (x, y) - (U^*x, y)$$

$$= (x - U^*x, y),$$

that is, that $x - U^*x = x - U^{-1}x$ is orthogonal to every vector y, so that $x - U^{-1}x = 0$, $x = U^{-1}x$, or $Ux = x$. Reading the last computation from right to left shows that this necessary condition is also sufficient; we need only to recall the definition of $\mathfrak{M}$ to see that $\mathfrak{M} = \mathfrak{N}^{\perp}$.

This very ingenious proof, which works with only very slight modifications in most of the important infinite-dimensional cases, is due to F. Riesz.

§ 93. Power series

We consider next the so-called Neumann series $\sum_{n=0}^{\infty} A^n$, where A is a linear transformation with norm <1 on a finite-dimensional vector space. If we write

$$S_p = \sum_{n=0}^{p} A^n,$$

then

(1) $$(1 - A)S_p = S_p - AS_p = 1 - A^{p+1}.$$

To prove that S_p has a limit as $p \to \infty$, we consider (for any two indices p and q with $p > q$)

$$\| S_p - S_q \| \leq \sum_{n=q+1}^{p} \| A^n \| \leq \sum_{n=q+1}^{p} \| A \|^n.$$

Since $\| A \| < 1$, the last written quantity approaches zero as $p, q \to \infty$; it follows that S_p has a limit S as $p \to \infty$. To evaluate the limit we observe that $1 - A$ is invertible. (Proof: $(1 - A)x = 0$ implies that $Ax = x$, and, if $x \neq 0$, this implies that $\| Ax \| = \| x \| > \| A \| \cdot \| x \|$, a contradiction.) Hence we may write (1) in the form

(2) $$S_p = (1 - A^{p+1})(1 - A)^{-1} = (1 - A)^{-1}(1 - A^{p+1});$$

since $A^{p+1} \to 0$ as $p \to \infty$, it follows that $S = (1 - A)^{-1}$.

As another example of an infinite series of transformations we consider the exponential series. For an arbitrary linear transformation A (not necessarily with $\| A \| < 1$) we write

$$S_p = \sum_{n=0}^{p} \frac{1}{n!} A^n.$$

Since we have

$$\| S_p - S_q \| \leq \sum_{n=q+1}^{p} \frac{1}{n!} \| A \|^n,$$

and since the right side of this inequality, being a part of the power series for $\exp \| A \| = e^{\|A\|}$, converges to 0 as $p, q \to \infty$, we see that there is a linear transformation S such that $S_p \to S$. We write $S = \exp A$; we shall merely mention some of the elementary properties of this function of A.

Consideration of the triangular forms of A and of S_p shows that the proper values of $\exp A$, together with their algebraic multiplicities, are equal to the exponentials of the proper values of A. (This argument, as well as some of the ones that follow, applies directly to the complex case only; the real case has to be deduced via complexification.) From the consideration of the triangular form it follows also that the determinant of $\exp A$, that is, $\prod_{i=1}^{N} \exp \lambda_i$, where $\lambda_1, \cdots, \lambda_N$ are the (not necessarily distinct) proper values of A, is the same as $\exp (\lambda_1 + \cdots + \lambda_N) = \exp$

(tr A). Since exp $\zeta \neq 0$, this shows, incidentally, that exp A is always invertible.

Considered as a function of linear transformations the exponential retains many of the simple properties of the ordinary numerical exponential function. Let us, for example, take any two *commutative* linear transformations A and B. Since exp $(A + B) -$ exp A exp B is the limit (as $p \to \infty$) of the expression

$$\sum_{n=0}^{p} \frac{1}{n!} (A + B)^n - \sum_{m=0}^{p} \frac{1}{m!} A^m \cdot \sum_{k=0}^{p} \frac{1}{k!} B^k$$

$$= \sum_{n=0}^{p} \frac{1}{n!} \sum_{j=0}^{n} \binom{n}{j} A^j B^{n-j} - \sum_{m=0}^{p} \sum_{k=0}^{p} \frac{1}{m!k!} A^m B^k,$$

we will have proved the multiplication rule for exponentials when we have proved that this expression converges to zero. $\left(\text{Here } \binom{n}{j} \text{ stands for the}\right.$ combinatorial coefficient $\left. \frac{n!}{j!(n-j)!}.\right)$ An easy verification yields the fact that for $k + m \leq p$ the product $A^m B^k$ occurs in both terms of the last written expression with coefficients that differ in sign only. The terms that do not cancel out are all in the subtrahend and are together equal to

$$\sum_m \sum_k \frac{1}{m!k!} A^m B^k,$$

the summation being extended over those values of m and k that are $\leq p$ and for which $m + k > p$. Since $m + k > p$ implies that at least one of the two integers m and k is greater than the integer part of $\frac{p}{2}$ (in symbols $[\frac{p}{2}]$), the norm of this remainder is dominated by

$$\sum_{m=0}^{\infty} \sum_{k=[\frac{p}{2}]}^{\infty} \frac{1}{m!k!} \| A \|^m \| B \|^k$$

$$+ \sum_{k=0}^{\infty} \sum_{m=[\frac{p}{2}]}^{\infty} \frac{1}{m!k!} \| A \|^m \| B \|^k$$

$$= \left(\sum_{m=0}^{\infty} \frac{1}{m!} \| A \|^m \right) \left(\sum_{k=[\frac{p}{2}]}^{\infty} \frac{1}{k!} \| B \|^k \right)$$

$$+ \left(\sum_{k=0}^{\infty} \frac{1}{k!} \| B \|^k \right) \left(\sum_{m=[\frac{p}{2}]}^{\infty} \frac{1}{m!} \| A \|^m \right)$$

$$= (\text{exp} \| A \|)\alpha_p + (\text{exp} \| B \|)\beta_p,$$

where $\alpha_p \to 0$ and $\beta_p \to 0$ as $p \to \infty$.

Similar methods serve to treat $f(A)$, where f is any function representable by a power series,

$$f(\zeta) = \sum_{n=0}^{\infty} \alpha_n \zeta^n,$$

and where $\| A \|$ is (strictly) smaller than the radius of convergence of the series. We leave it to the reader to verify that the functional calculus we are here hinting at is consistent with the functional calculus for normal transformations. Thus, for example, exp A as defined above is the same linear transformation as is defined by our previous notion of exp A in case A is normal.

EXERCISES

1. Give an alternative proof of the ergodic theorem, based on the spectra theorem for unitary transformations.

2. Prove that if $\| 1 - A \| < 1$, then A is invertible, by considering the formal power series expansion of $(1 - (1 - A))^{-1}$.

APPENDIX

HILBERT SPACE

Probably the most useful and certainly the best developed generalization of the theory of finite-dimensional inner product spaces is the theory of Hilbert space. Without going into details and entirely without proofs we shall now attempt to indicate how this generalization proceeds and what are the main difficulties that have to be overcome.

The definition of Hilbert space is easy: it is an inner product space satisfying one extra condition. That this condition (namely, completeness) is automatically satisfied in the finite-dimensional case is proved in elementary analysis. In the infinite-dimensional case it may be possible that for a sequence (x_n) of vectors $\| x_n - x_m \| \to 0$ as $n, m \to \infty$, but still there is no vector x for which $\| x_n - x \| \to 0$; the only effective way of ruling out this possibility is explicitly to assume its opposite. In other words: a Hilbert space is a complete inner product space. (Sometimes the concept of Hilbert space is restricted by additional conditions, whose purpose is to limit the size of the space from both above and below. The most usual conditions require that the space be infinite-dimensional and separable. In recent years, ever since the realization that such additional restrictions do not pay for themselves in results, it has become customary to use "Hilbert space" for the concept we defined.)

It is easy to see that the space $\mathcal{O}$ of polynomials with the inner product defined by $(x, y) = \int_0^1 x(t)\overline{y(t)}\, dt$ is not complete. In connection with the completeness of certain particular Hilbert spaces there is quite an extensive mathematical lore. Thus, for instance, the main assertion of the celebrated Riesz-Fischer theorem is that a Hilbert space manufactured out of the set of all those functions x for which $\int_0^1 |x(t)|^2\, dt < \infty$ (in the sense of Lebesgue integration) is a Hilbert space (with formally the same definition of inner product as for polynomials). Another popular Hilbert space, reminiscent in its appearance of finite-dimensional coordinate space, is the space of all those sequences (ξ_n) of numbers (real or complex, as the case may be) for which $\sum_n |\xi_n|^2$ converges.

Using completeness in order to discuss intelligently the convergence of some infinite sums, one can proceed for quite some time in building the theory of Hilbert spaces without meeting any difficulties due to infinite-dimensionality. Thus, for instance, the notions of orthogonality and of complete orthonormal sets can be defined in the general case exactly as we defined them. Our proof of Bessel's inequality and of the equivalence of the various possible formulations of completeness for orthonormal sets have to undergo slight verbal changes only. (The convergence of the various infinite sums that enter is an automatic consequence of Bessel's inequality.) Our proof of Schwarz's inequality is valid, as it stands, in the most general case. Finally, the proof of the existence of complete orthonormal sets parallels closely the proof in the finite case. In the unconstructive proof Zorn's lemma (or transfinite induction) replaces ordinary induction, and even the constructive steps of the Gram-Schmidt process are easily carried out.

In the discussion of manifolds, functionals, and transformations the situation becomes uncomfortable if we do not make a concession to the topology of Hilbert space. Good generalizations of all our statements for the finite-dimensional case can be proved if we consider *closed* linear manifolds, *continuous* linear functionals, and *bounded* linear transformations. (In a finite-dimensional space every linear manifold is closed, every linear functional is continuous, and every linear transformation is bounded.) If, however, we do agree to make these concessions, then once more we can coast on our finite-dimensional proofs without any change most of the time, and with only the insertion of an occasional ϵ the rest of the time. Thus once more we obtain that $\mathcal{U} = \mathfrak{M} \oplus \mathfrak{M}^\perp$, that $\mathfrak{M} = \mathfrak{M}^{\perp\perp}$, and that every linear functional of x has the form (x, y); our definitions of self-adjoint and of positive transformations still make sense, and all our theorems about perpendicular projections (as well as their proofs) carry over without change.

The first hint of how things can go wrong comes from the study of orthogonal and unitary transformations. We still call a transformation U orthogonal or unitary (according as the space is real or complex) if $UU^* = U^*U = 1$, and it is still true that such a transformation is isometric, that is, that $\| Ux \| = \| x \|$ for all x, or, equivalently, $(Ux, Uy) = (x, y)$ for all x and y. It is, however, easy to construct an isometric transformation that is not unitary; because of its importance in the construction of counterexamples we shall describe one such transformation. We consider a Hilbert space in which there is a countable complete orthonormal set, say $\{x_0, x_1, x_2, \cdots\}$. A unique bounded linear transformation U is defined by the conditions $Ux_n = x_{n+1}$ for $n = 0, 1, 2, \cdots$. This U is isometric ($U^*U = 1$), but, since $UU^*x_0 = 0$, it is not true that $UU^* = 1$.

It is when we come to spectral theory that the whole flavor of the development changes radically. The definition of proper value as a number λ for which $Ax = \lambda x$ has a non-zero solution still makes sense, and our theorem about the reality of the proper values of a self-adjoint transformation is still true. The notion of proper value loses, however, much of its significance. Proper values are so very useful in the finite-dimensional case because they are a handy way of describing the fact that something goes wrong with the inverse of $A - \lambda$, and the only thing that can go wrong is that the inverse refuses to exist. Essentially different things can happen in the infinite-dimensional case; just to illustrate the possibilities, we mention, for example, that the inverse of $A - \lambda$ may exist but be unbounded. That there is no useful generalization of determinant, and hence of the characteristic equation, is the least of our worries. The whole theory has, in fact, attained its full beauty and maturity only after the slavish imitation of such finite-dimensional methods was given up.

After some appreciation of the fact that the infinite-dimensional case has to overcome great difficulties, it comes as a pleasant surprise that the spectral theorem for self-adjoint transformations (and, in the complex case, even for normal ones) does have a very beautiful and powerful generalization. (Although we describe the theorem for bounded transformations only, there is a large class of unbounded ones for which it is valid.) In order to be able to understand the analogy, let us re-examine the finite-dimensional case.

Let A be a self-adjoint linear transformation on a finite-dimensional inner product space, and let $A = \sum_j \lambda_j F_j$ be its spectral form. If M is an interval in the real axis, we write $E(M)$ for the sum of all those F_j for which λ_j belongs to M. It is clear that $E(M)$ is a perpendicular projection for each M. The following properties of the projection-valued interval-function E are the crucial ones: if M is the union of a countable collection $\{M_n\}$ of disjoint intervals, then

(1) $$E(M) = \sum_n E(M_n),$$

and if M is the improper interval consisting of all real numbers, then $E(M) = 1$. The relation between A and E is described by the equation

$$A = \sum_j \lambda_j E(\{\lambda_j\}),$$

where, of course, $\{\lambda_j\}$ is the degenerate interval consisting of the single number λ_j. Those familiar with Lebesgue-Stieltjes integration will recognize the last written sum as a typical approximating sum to an integral of the form $\int \lambda \, dE(\lambda)$ and will therefore see how one may expect the generalization to go. The algebraic concept of summation is to be replaced by the analytic concept of integration; the generalized relation between A and E is described by the equation

$$(2) \qquad A = \int \lambda \, dE(\lambda).$$

Except for this formal alteration, the spectral theorem for self-adjoint transformations is true in Hilbert space. We have, of course, to interpret correctly the meaning of the limiting operations involved in (1) and (2). Once more we are faced with the three possibilities mentioned in § 91. They are called uniform, strong, and weak convergence respectively, and it turns out that both (1) and (2) may be given the strong interpretation. (The reader deduces, of course, from our language that in an infinite-dimensional Hilbert space the three possibilities are indeed distinct.)

We have seen that the projections F_j entering into the spectral form of A in the finite-dimensional case are very simple functions of A (§ 82). Since the $E(M)$ are obtained from the F_j by summation, they also are functions of A, and it is quite easy to describe what functions. We write $g_M(\zeta) = 1$ if ζ is in M and $g_M(\zeta) = 0$ otherwise; then $E(M) = g_M(A)$. This fact gives the main clue to a possible proof of the general spectral theorem. The usual process is to discuss the functional calculus for polynomials, and, by limiting processes, to extend it to a class of functions that includes all the functions g_M. Once this is done, we may define the interval-function E by writing $E(M) = g_M(A)$; there is no particular difficulty in establishing that E and A satisfy (1) and (2).

After the spectral theorem is proved, it is easy to deduce from it the ge neralized versions of our theorems concerning square roots, the functional ca lculus, the polar decomposition, and properties of commutativity, and, in fact, to answer practically every askable question about bounded normal tr ansformations.

The chief difficulties that remain are the considerations of non-normal an d of unbounded transformations. Concerning general non-normal transformations, it is quite easy to describe the state of our knowledge; it is non-existent. No even unsatisfactory generalization exists for the triangular form or for the Jordan canonical form and the theory of elementary divisors. Very different is the situation concerning normal (and particularly self-adjoint) unbounded transformations. (The reader will sympathize with the desire to treat such transformations if he recalls that the first and most important functional operation that most of us learn is differentiation.) In this connection we shall barely hint at the main obstacle the theory faces. It is not very difficult to show that if a self-adjoint linear transformation is defined for all vectors of Hilbert space, then it is bounded. In other words, the first requirement concerning transformations that we are forced to give up is that they be defined everywhere. The discussion of the precise domain on which a self-adjoint transformation may be defined and of the extent to which this domain may be enlarged is the chief new difficulty encountered in the study of unbounded transformations.

RECOMMENDED READING

The following very short list makes no pretense to completeness; it merely contains a couple of representatives of each of several directions in which the reader may want to proceed.

For generalized (but usually finite-dimensional) linear and multilinear algebra:

1. N. BOURBAKI, *Algèbre;* Chap. II (*Algèbre linéaire*), Paris, 1947, and Chap. III (*Algèbre multilinéaire*), Paris, 1948.
2. B. L. VAN DER WAERDEN, *Modern algebra*, New York, 1953.

For connections with classical and modern analysis:

1. S. BANACH, *Théorie des opérations linéaires*, Warszawa, 1932.
2. F. RIESZ and B. SZ.-NAGY, *Functional analysis*, New York, 1955.

For the geometry of Hilbert space and transformations on it:

1. P. R. HALMOS, *Introduction to Hilbert space*, New York, 1951.
2. M. H. STONE, *Linear transformations in Hilbert space*, New York, 1932.

For contact with classical and modern physics:

1. R. COURANT and D. HILBERT, *Methods of mathematical physics*, New York, 1953.
2. J. VON NEUMANN, *Mathematical foundations of quantum mechanics*, Princeton, 1955.

INDEX OF TERMS

INDEX OF SYMBOLS

Printed in the United States
28438LVS00002BB/61

9 780387 900933